CALIFORNIA NATURAL HISTORY GUIDES

FIELD GUIDE TO THE
COMMON BEES OF CALIFORNIA

California Natural History Guides

Phyllis M. Faber, General Editor

www.uidai.gov.in

Unique Identification Authority of India

आधार

Address: Attari, Attari, Amritsar, Punjab,
143108

4217 6668 2926

help@uidai.gov.in

1947

Government of India

Mohani

DOB : 01/05/2004

Female

4217 6668 2926

मेरा आधार, मेरी पहचान

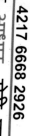

The publisher gratefully acknowledges the
generous contributions to this book provided by
the Gordon and Betty Moore Fund
in Environmental Studies.

———————

CONTENTS

PREFACE

Why a Guide to Bees?

From childhood on, the sight of bees visiting flowers is part of our experience of the natural world. We marvel at their industry and the mystery of their travels from flower to flower and flower to hive. The scientific study of bees has revealed that bees are an essential part of healthy ecosystems. Many plants, including important crop plants, depend on the pollination they provide—by carrying the pollen that bees use for food from flower to flower, bees ensure that flowers receive ample pollen for fertilization and seed set. There is growing evidence that many populations of bees are declining, and conservation of bees and other pollinators is a growing concern. Using this guide to identify some of the common groups of bees in California is a satisfying challenge, and it is my hope that the study of bees at any level can provide an excellent point of departure into the study of the bond between plants and pollinators and involvement in pollinator conservation.

This guide is meant to introduce the nonprofessional to the fascinating biology of bees and to give you some tools for beginning to know the key genera of bees. By learning about bees at this level, you can know a tremendous amount about the particular biology of any bee you catch, since bees in the same genera often share nesting habits and are often similar in their degree of specialization on flowers. While we have focused on California bees, the characteristics of the genera included in this introduction can be applied across the United States and Canada in most cases. We do not provide a guide to the species of bees except for the Guide to Some of the More Common Bumble Bees of California. Bees are extraordinarily difficult to identify to the level of species, and even for scientists who specialize in knowing bees, identifying our California bees to the species level is a challenge. If you are interested in learning more about identification, we have provided a list of Key Readings and Key Web Resources in the Resources section at the end of the book.

How to Use This Book

This book is directed at helping you learn about the common genera of California bees. I have chosen to introduce the genera of bees because bees are significantly harder to identify at both the family level and the species level. The taxonomy follows *The Bees of the World* by Charles Michener (2007). While many species have very restricted ranges, most of the genera covered in this book are found across all of California, and many across the whole United States. I have, therefore, chosen not to include distribution maps. Excellent distribution maps for California species can be found on the website for Discover Life at www.discover life.org.

I have provided two guides in this book. In the Bee Family and Genus Accounts, you will find a Key of Basic Bee Characteristics that will allow you to distinguish a few genera with easily discernible traits. Then, at the end of the book, you will find Appendix 2, Key to Females of Each Genus Included in the Book, which provides a matrix for distinguishing among the genera. Using it, you can fairly quickly narrow down the genera by simply determining whether the bee is larger or smaller than a worker honey bee, the bee's color, and whether the bee has external patterning. Then, you can look at the genus account to narrow it down further. In many cases, the technical description in the genus account will require that you look at the bee under magnification to see the critical characteristics that are used to identify that genus of bees.

Acknowledgments

I would like to thank Jerry Rozen, Rob Brooks, Robbin Thorp, Terry Griswold, Charles Michener, Jim Cane, Steve Buchmann, Bob Minckley, James Thomson, Laurence Packer, Bryan Danforth, and the many students of the Southwestern Research Station's Bee Course for all they have taught us about bees. The genus accounts were developed using the resources on the Discover Life website, which owes a huge debt to John Ascher. Additional recognition goes to John Ascher, who has compiled and made available so much information about the distribution and taxonomy of bees. Special recognition goes to Sam Droege for all he has done to make collecting bees easier and for contributing the key to bees in Appendix 2. The pronunciations used in this guide were developed by Sam Droege. We would also like to recognize the major contributions of Mace Vaughan and Lisa Schonberg of the Xerces Society, who helped draft early versions of some of the genus accounts. Lynn Lozier and Mike Mesler provided excellent advice on how to improve the manuscript. Leah Larkin, Fred Bove, and Mark Reynolds helped with editing of the manuscript. Much of the information on California genera came from the magnificent catalog of California bees by Andy Moldenke and Jack Neff.

AN INTRODUCTION TO BEES

What Are Bees?

Anyone who has spent a lovely warm morning in a garden in spring has shared company with a bee. From a farm in the Central Valley to a community garden in the center of Los Angeles, bees are busy buzzing around, visiting flowers, gathering resources for their offspring, and in the process, transferring pollen from flower to flower.

Bees are flying insects that first emerged about 100 million years ago, during the Cretaceous period, as part of the radiation of insects. The earliest record of bees is from fossilized amber in Myanmar (formerly Burma).

Bees are part of a larger group of insects, the order Hymenoptera that includes wasps, sawflies (a primitive and less well-known group), and ants. Bees belong to a large group within the Hymenoptera that are distinguished, in part, by having females with stingers. This group is called the Aculeata and contains the bees, ants, and stinging wasps. Recent work on the relationship between bees and wasps shows that bees are really just a very specialized group of wasps. Formally, we say that bees are in the superfamily Apoidea along with the sphecid wasps and make up the group Apiformes. No wonder bees and wasps can be so hard to tell apart!

While beetles were probably the first pollinators, bees were the first group of insects to really show a diversity of adaptations for pollination, such as specialized hairs and leg modifications, both for carrying pollen. This close relationship between bees and flowers is thought to have aided in the rapid evolution of different types of flowering plants and has strongly influenced both the body plans and lifestyles of bees and flowers.

When you look at most bees, you see a fairly hearty and hairy flying insect. Although there are bees that are slender and sleek, the most well-known, like the bumble bees and honey bees, are robust and hairy. Their hairs are an essential part of the service that bees are renowned for, transferring pollen from plant to plant. The hairs hold pollen onto the bee body. However, while it may seem bees are providing this service to plants free of charge—almost as if the flowers were taking advantage of them—the bees are actually getting great benefits from the interaction. What the bees are really doing is searching out food for their offspring and themselves. Bees rely on plant pollen for protein and nectar for energy. In fact, the sole source of protein for most bees is pollen. These bees could not exist without flowers to feed them.

When you see a bee visiting a flower, it is probably either collecting pollen or drinking nectar. If the bee is collecting pollen at flowers, she is a female gathering food to take back to her nest for her larvae. She might stop to fill up on some nectar, but her primary aim is to gather pollen for her offspring. This behavior highlights one big difference between bees and the closest wasp relatives of bees, the sphecoid wasps: the larvae of sphecoids are carnivorous, eating spiders and insects, whereas bee larvae are vegetarians, relying on pollen for protein. Of course, as with everything in nature, there is an exception to this rule: a group of tropical bees that feed their offspring carrion. If you see a bee simply drinking nectar, it is more likely to be a male bee. Male bees do visit flowers to tank up on nectar, but they do not collect pollen, as they do not provision nests.

At first glance, it can be difficult to tell bees, flies, and wasps apart because they have similar sizes and colors. Flies and wasps are confusing because they often look like bees. Some striped flies may actually be bee mimics, trying to fool predators! To tell bees and wasps from flies, there are three main distinguishing features: the number of wings that there are on each side, the shape of the antennae, and where the eyes are placed on the face. Let's start by distinguishing flies from bees. First, the number of wings on a side makes it easy to distinguish flies from bees and wasps. Flies have a single wing on each side. Bees and wasps have two per side.

MIMICRY

Many distasteful or poisonous species that are trying to signal a warning use combinations of the colors yellow, red, and black. From snakes to butterflies to bees, those colors make them very apparent and protect them from attack. Predators quickly learn to avoid prey with particular color patterns. The color patterns of bumble bees are a good example of warning coloration. When two or more poisonous species share a predator and have a similar color pattern that cannot be attributed to related-ness, scientists call it Müllerian mimicry. This might explain the similarity in color patterns between bees and wasps. Another explanation for the similarity of color patterns might be that they all inherited the pattern from a common ancestor. We also see a different kind of mimicry of bees. When a harmless species mimics the color pattern of a dangerous species, it is called Batesian mimicry. In this situation, the dangerous species is called the model and the common species is the mimic. A number of species of flies have color patterns very similar to those of bees, especially bumble bees. These are probably examples of Batesian mimicry. For Batesian mimicry to be effective and maintained, the model must be more frequently encoun-tered than the mimic. Otherwise, the predator may learn that the mimic is a good prey item and periodically consume some of the model.

It can be hard to tell that bees have a pair of wings on each side, because the wings on bees and wasps can be hooked together with special hooks called hamuli. When looking for bees on flowers, you can often tell that an insect is a fly because its wings are not neatly folded over its back; instead they point out at an angle. Second, you can look at the insect's antennae. Flies generally have short, thick antennae, whereas bees have longer, thinner antennae. Finally, bees have large eyes that are on the side of the head. Flies have large eyes on the front of the head. Other general rules of thumb are that bees tend to be hairy, and they carry loads of pollen, whereas flies have fewer hairs and do not generally carry pollen. This does not work when distinguishing male bees from flies, as male bees do not generally carry pollen and do not have elaborate structures like scopae, the pollen-carrying structures on a bee, for doing so. Behaviorally, bees do not hover whereas many but not all flies do.

Wasps are much more difficult to distinguish from bees. Given that bees and wasps are much more closely related than bees and flies, this is not too surprising. Bees and wasps both have two pairs of wings per side and are often the same size, shape, and color. In general, wasps look "meaner" than bees. This may be because wasps seem to have more armoring in their exoskeleton, the protective covering over the body, than bees. Bees tend to have a broader body and wider abdomen than wasps. Bees' bodies are usually hairy, and in general, their exoskeleton is a single color, except for stripes on the abdomen. Wasps are less hairy and often have patterns or designs in their exoskeleton. One unique thing about wasps is that the hairs on their faces are often shiny or metallic, often silvery, whereas bees have duller hairs or no hairs. Under magnification, bees can be identified by the presence of plumose or branched hairs on some parts of their body and their legs and by the hind basitarsus (basal segment of the tarsus), which is often more flattened (broader in one dimension and narrower in another) than the next segments of the tarsus.

Importance of Bees

The conservation of bees is central to biodiversity and life on Earth, to food security, and to the global economy. While bees are well known for their honey production, it is pollination that makes them critical to life on Earth. Bees are involved in pollinating about 70 percent of the world's flowering plants. For flowering plants, which cannot move themselves, having a bee move pollen dramatically increases the distance pollen can travel to fertilize seeds and thereby increases the potential number of mates. There are many advantages to expanding the number of potential mates. In particular, mating with more individuals increases genetic variation, which provides the raw material for evolution and adaptation, and those populations with greater genetic variation are going to have a higher probability of surviving as environments change.

There are several examples of plants that seem to be declining more than they would if they had a healthy pollinator community. For example, in the Antioch Dunes of the San Francisco Bay Area, there is a primrose *(Oenothera deltoides* subsp. *howellii)* that is pollinated both by hawk moths and by bees. In fact, it has a specialist bee that visits it, *Sphecodogastra antiochensis,* that flies in the early morning and evening when the flowers are open, and this bee is now rare. This primrose is producing only 35 percent of the seeds that it might with a healthy pollinator community. Both the specialist bee and the primrose are considered rare or endangered and are only found within the 75 remaining acres of the Antioch Dunes.

If we think about the benefits of bees to people, the pollination of food plants is equally important. Bees pollinate about 75 percent of the fruit, nuts, and vegetables grown in California. Amazingly, with this much of the food supply at stake, agriculture relies almost exclusively on a single pollinator, the Western Honey Bee, *Apis mellifera.* In the United States alone, the value to crop production of pollination by this single species is estimated to be $14 billion. Our reliance on this species is still increasing. Demand for Honey Bee colonies increased approximately 25 percent from 1989 to 1998 and has continued to increase since that time.

Pollinator service to plants depends on both Honey Bees and other, native pollinators. We know native bees can have a significant role, because increases in the number and species of native bees have been associated with increases in crop production. The actual value of native pollinators for crop production is much more difficult to estimate, because the values of specific pollinators differ for each crop and are dependent on geographic location, availability of natural habitat, and use of pesticides. Studies in agricultural systems have shown that native pollinators can provide significant pollinator service directly, by pollinating the plants themselves, and also indirectly because the efficiency of Honey Bees is improved by the presence of native bees. For example, Sarah Greenleaf and Claire Kremen found that when native bees were foraging in sunflower fields in the Central Valley, Honey Bee efficiency at pollination increased up to fivefold. This increase in efficiency is because Honey Bees move more often when in the presence of native bees, perhaps because of competition.

In addition to their role as pollinators of crops, animals pollinate approximately three-fourths of the flowering plants (angiosperms). Declines in pollinators may increase the risk of extinction of many native plant species, as we are seeing for the Antioch Dunes Evening Primrose, particularly in biodiversity hotspots like the Bay Area. In these systems, bee pollinators other than Honey Bees are of primary importance.

Bee Diversity

When I talk about bees, most people immediately assume that I am talking about Western Honey Bees. In fact, Honey Bees are only one of the approximately 20,000 species of bees in the world. Here in North America, scientists estimate there are about 4,000 species, and almost 1,500 species, or almost 40 percent of those North American bee species, are found in California. The bees in California range in size from a tiny sweat bee *(Perdita)*, not much larger than the head of a pin, to a large carpenter bee *(Xylocopa)*, the size of a man's thumb. There are 82 genera of bees known from California. Many are represented by a small number of species. I have chosen to profile the most common genera across California. See Appendix 1 for a complete list of California bee families and genera.

Bees come in a variety of shapes, sizes, and colors. They nest in a variety of places, including holes in trees and underground. Bees can be social, living in large family groups with different roles for different bees, or they can be solitary, with a single queen working and nesting alone. The surprising diversity in bees is part of their fascination to humans. Over the next pages, I will try to share some of that wonder.

Bee Morphology

To understand bees, it helps to know how they are put together. There are three main parts to a bee. These are usually called the head, the thorax, and the abdomen (Figure 1). The word *thorax* is a bit misleading for bees because the final segment attached to this part of the bee's body is actually part of the abdomen in other insects. This structure is the propodeum. The eight segments of the abdomen are each encased in two plates: the upper plate of a segment is called a tergum (plural terga), and the bottom plate is the sternum (plural sterna). A lot of bees have pale bands of hair across their terga. The number of terga and sterna are important for distinguishing male and female bees. Male bees have seven and female bees have six. If you look at the end of the abdomen, you may see a fringe of hairs and a small plate called the pygidial plate. This plate is used to tamp down soil within burrows and nest cells. When you are looking at

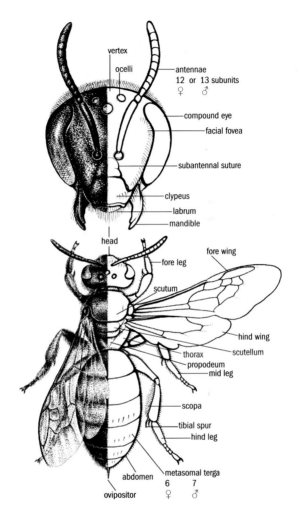

Figure 1

a female bee, you may also see an ovipositor. The ovipositor is the structure that a female uses to lay her eggs. The stinger is derived from the ovipositor and has a less pleasant function in defense. In many species, the female bee is larger than the male.

In many California native bees, the stinger is not strong enough to penetrate human skin. However, some bees pack a wallop. When stinging, the stinger is pushed out from the abdomen, and then it locks into position. In a situation where a bee is stinging, it will use the muscular plates on the abdomen to push the stinger into your skin. The stinger has a very sharp tip, and at the top of the stinger there is a venom bulb. In the Western Honey Bee, the stinger breaks off from the abdomen, yet the venom bulb will continue to pump venom for 30–60 seconds. While Honey Bee workers die after stinging, this is not true of all bees.

The abdomen and thorax of bees house most of their internal organs. The heart, the stomach, air sacs for breathing, and the sex organs are all contained in these parts of the body. Bees breathe through small holes in their exoskeleton, called spiracles. These spiracles connect to air sacs and allow for the free passage of air. In addition, the thorax of the bees is where the legs and wings connect. The thorax is a very complex structure with numerous fused plates and sites where muscles attach.

Bees and other insects have six legs: two fore legs, two mid legs, and two hind legs. Each leg is made up of a coxa, trochanter, femur, and tibia tarsus and ends with a pre-tarsus (with claws). Legs are important for grooming, for courtship, for pollen collection and carrying, for constructing nests, and of course, for locomotion. When a female bee collects pollen, she usually gathers it with the fore legs, transfers it to the mid legs, and then stores in on the hind legs.

There are some unique functions that legs perform that correspond to their shapes. If you look at the apex of the tibia, you can often see a small spike called a tibial spur. The spurs on the fore leg are modified into a tool that bees use to clean antennae. We do not know the function of the tibial spurs on the other legs. You can actually watch bees groom their antennae with their fore legs. They run the antennae along the intersection of the tibial spur and the leg itself. On the hind legs of many bees, you can see scopae, or pollen-collecting hairs where pollen is concentrated before the trip back to the nest. The scopa is a complex structure made of modified hairs on the coxa, trochanter, femur, tibia, or underside of the abdomen (in the family Megachilidae) and even the sides of the propodeum (in some members of the family Halictidae and the genus *Andrena*). There is more about pollen-carrying structures below.

Bee wings are composed of a transparent wing membrane supported by a stiff network of round wing veins (Figure 2). The patterns of wing veins are very variable and are really important for identifying groups of bees. The veins add strength to the wings. There are two wings on each side of a bee: a fore wing and a hind wing. The leading edge of the hind wing has little hooks on it called hamuli (Figure 3). These hooks hold the fore wing and hind wing together in flight. Hamuli stabilize the wings and influence the bee's ability to fly. The number of hamuli on a wing is related to the size of the bee and how far it flies. Wings beat about 400–500 times per minute. Amazingly, it has been documented that there are some tropical bees that fly over 14 miles in a single trip.

You may have heard that from an engineering standpoint, bumble bees should not be able to fly. While the originator of the calculation that shows that it should be impossible for a bumble bee to get off the ground remains in dispute (attributions range from a French entomologist to a German physicist), the same error is attributed to each. Whoever it was, he estimated the amount of lift that a bumble bee would have, given its body and wing sizes. To keep the calculations simple, he used a model for a rigid, smooth, fixed-wing

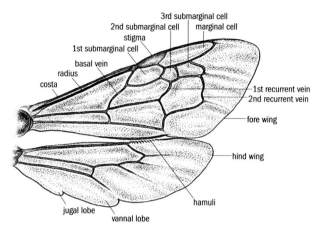

Figure 2

craft. From these calculations, it appeared that a bumble bee would have insufficient lift to take flight. In fact, bumble bee wings are very flexible and have considerable movement and would be better modeled using something more similar to a helicopter.

The most obvious features on the head of a bee are the large compound eyes, the three simple eyes called ocelli, and the antennae. The compound eyes are made up of over 6,000 tiny lenses. Each lens is connected to a tube that has receptors that respond to polarized light. Interestingly, research on Western Honey Bees has shown that when a bee leaves the nest, she will fly around in a circle until her eyes receive maximum stimulation from the polarized light. This signal tells her that she is facing away from the sun and allows her to orient her flight. However, a bee's vision is believed to be sharp for a distance of only about 3 ft. Bees that fly in the early morning, late evening, or night have enlarged ocelli. For these bees flying in low-light conditions, the ocelli might function as light receptors that help them keep their bodies level while flying. The antennae are used

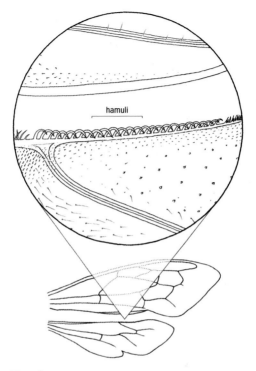

Figure 3

both to touch and to sense the environment in other ways. When you look at the face of a bee, using a microscope or a powerful hand lens, notice there are two sutures running down from the antennae. These are the subantennal sutures (see Figure 1). They are important features for identification because if there are two subantennal sutures, you can know it is a bee in the family Andrenidae. Most other bees have just one suture below each eye. With a microscope, you can also see depressions in the face called facial foveae that indicate the site of some glandular structures. They are found along the inner margins of the compound eyes in many bees, particularly in the families Andrenidae and Colletidae.

Now, if you look underneath the face, you will see the lower face and mouthparts (Figure 4). The mouthparts sit within a deep groove in the head. The lower face and mouthparts are made up of a pair of mandibles plus the labrum. Mandibles are used to bite, work wax and pollen, and carry objects such as resins and mud. The clypeus is the plate on the front of the head of a bee, below the antennae. The labrum is attached to the bottom of the clypeus. Bees are often divided into two groups: the long-tongued bees and the short-tongued bees. The short-tongued bees (the families Colletidae, Andrenidae, Halictidae, and Melittidae) have labial palpi that are all similar in size and shape, whereas in the long-tongued bees (the families Apidae and Megachilidae) the first two labial palpi are long, flat, and almost like a sheath (see Figure 4).

If you do not have a bee under magnification, the most obvious way to tell a male bee from a female bee is by whether it is carrying pollen or has pollen-carrying features like scopae (Figure 5). Females do carry pollen and males generally do not. This is generally true, though in some bees, such as the genus *Hylaeus* and the parasitic bees, females do not have external pollen-carrying features, and males sometimes inadvertently get pollen on them. There are some other characteristics that are less reliable. Males often have relatively longer, narrower bodies; less-hairy legs; and white markings or hairs on their faces. There are some groups, particularly the families Apidae and Halictidae, where males have much longer antennae. In a few species, notably the carpenter bees (genus *Xylocopa*) and some bumble bees (genus *Bombus*), males and females have different color patterns.

Using a microscope, it is fairly simple to tell male and female bees apart. Male bees have antennae with 13 subunits

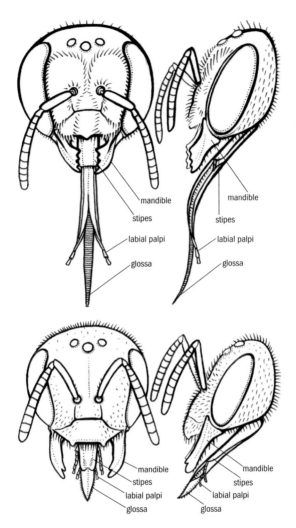

mandible
stipes
labial palpi
glossa

mandible
stipes
labial palpi
glossa

mandible
stipes
labial palpi
glossa

mandible
stipes
labial palpi
glossa

Figure 4

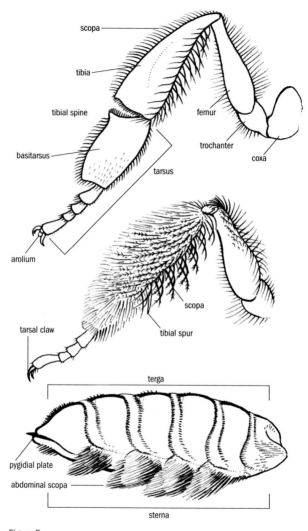

scopa

tibia

tibial spine

basitarsus

arolium

tarsus

femur

trochanter

coxa

tarsal claw

scopa

tibial spur

terga

pygidial plate

abdominal scopa

sterna

Figure 5

(see Figure 1); females have 12 subunits. Also, males have seven exposed metasomal terga; females have six. Even without a microscope, you can often see the stinger on a female bee.

Bee Life Cycles

All bees go through a series of developmental stages and follow a pattern of metamorphosis. Bees are considered holometabolous, which means that they follow a set of stages of development similar to those of butterflies and beetles. A holometabolous insect starts as an egg. The egg hatches into a larva (e.g., a caterpillar, grub, maggot; plural is larvae), which goes through an inactive, pupal stage (e.g., wrapped up like a cocoon; singular is pupa, plural is pupae) before emerging as an adult (e.g., a butterfly, beetle, wasp, or bee). In most of our solitary bees, an adult female emerges in spring or summer. Her emergence is usually tied to the timing of the plants she needs in order to provision the nest cells for her offspring. The actual cues for emergence are not well understood, though they are suspected to be related to air temperatures or carbon dioxide levels in the soil. This female bee will spend all of her time building and provisioning a nest or nests filled with brood chambers. In each brood chamber, she will lay a single egg. Most bees are active as adults only for 2–5 weeks. They spend the other 47–50 weeks of the year in their nests in cavities or underground.

The female bee lays her egg on provisions. These provisions are made up of nectar and pollen. Many provisions look like a round ball but there are some that are very soupy. The quality and quantity of pollen and the amount of sugar in the nectar in those provisions affect the growth of larvae. Larvae with provisions made of higher-quality pollen and with higher sugar content grow larger and faster. Interestingly, there is no evidence that specialist bees use pollen with higher protein contents. Specialist bees also do not develop well on pollens from plants they do not normally use.

There are several different interesting ways that male bees find females. Some male bees set up territories by marking flowers with pheromones, which are scents for attracting females. Males in the genera *Andrena, Nomada,* and *Triepeolus* do not just mark flowers; they also set up "patrol routes" by marking nonflowering plants. Other males are often attracted to those scents and join the patrol. When a female bee enters the patrol route, a first male will attempt to mate with her. If there is more than one male in the area, the female will be mobbed. Larger mobs can actually tumble to the ground in the excitement. In some bees, the males patrol nest sites before females emerge, so they can be the first to mate. Males in other genera set up territories that they defend from other males.

The males of some species are aggressive. These males either patrol their territory (as some *Megachile* do) or defend a single site (as with *Xylocopa*). The size of male territories can be as small as a single flower clump or much larger (up to 20 sq. ft.). Territories probably change owners several times during breeding season, and the desirability of the territory probably changes as the floral resources and amount of competition change with the season.

Mating can take place in flight over the nesting site, inside a ground nest, or at or in a flower or simply on the ground (Figure 6). Males of many species can be found patrolling quickly through patches of flowers searching for females. Often, the mating pair is mobbed by other males.

Bumble bees actually use a queen-attracting scent to find mates. A male bee will fly in a circuit, putting scent (pheromone) in suitable places such as tree trunks or rocks. Different species of bumble bees have been shown to have different preferences for the height of the scent, ranging from low on bushes to high in tree tops. The scents themselves are different. There are even some scents that are detectable by human noses. New bumble bee queens emerge about a week after the males. When the new queen is ready to mate,

Figure 6

she will locate a site with pheromone and wait for a suitable mate. Mating usually takes place on the ground or on vegetation. Periodically you may see a large queen flying with a small male attached to her—still mating.

Matings usually last 10–80 minutes. The sperm transfer happens right at the start, often within 2 minutes. So why does the mating take so long? After the sperm has been transferred, the male transfers some sticky material that will harden in the female's genital opening. This genital plug blocks the entry of sperm from other males. Therefore, it is in the best interest of the male to continue mating so the plug has time to harden, thereby increasing the probability of his genes getting passed on to the next generation. Most bumble bee queens have only one mate, though some species do mate multiple times.

After a bumble bee queen has mated, she will find a spot to overwinter. Usually these spots are in shallow holes in the ground, often under a root or on a gentle slope. This overwintering site is called a hibernaculum (plural hibernacula). During the time from her emergence to her entry into the hibernaculum, the queen will visit many flowers to drink nectar and build up fat supplies in her body. She also fills her honey stomach. Bumble bees have a large honey stomach. This stomach is located in the abdomen and can hold 0.06–0.20 ml, depending on the size of the bumble bee. When full, the honey stomach can take up as much as 95 percent of the abdominal space and account for as much as 90 percent of the body weight. The body fat and honey stomach resources allow the queen to hibernate. In areas where it freezes, if the temperature in the ground gets very cold, the queen's body will start to produce glycerol, which prevents ice crystals from forming in her cells. If she didn't produce glycerol, when the fluids froze, they could expand to the point that they might burst the cells. After the queen emerges in spring, she will not reuse the hibernaculum for her nest site.

Evolution of Social Behavior: Social versus Solitary Bees

When most people think of bee societies, they think of Western Honey Bees with their queen and caste system of workers. Most North American bees do not have castes of workers like Honey Bees. Rather than being social, they are solitary, meaning that a single female works on her nest by herself. However, several other groups of bees are social. There is a set of formal criteria that defines sociality for animals. To be considered social in the scientific sense of the term, a bee must share the work of caring for the brood, there must be an overlap in generations so that offspring can assist parents, and there must be a reproductive division of labor, which means that not every female produces offspring. Among the social bees, there are different degrees of sociality. The most complex relationships are the eusocial bees like Honey Bees. These bees share a nest, have cooperative brood care and worker castes, and have overlapping generations. Between the eusocial bees and the solitary bees, there are several intermediate levels of interaction. Semisocial bees have small colonies of bees that are of the same generation. These bees share a nest and have worker castes and cooperative brood care. Usually one of the sisters becomes the queen and does most of the egg laying. Some of our bees in the family Halictidae exhibit this type of sociality. Communal bees nest together but do not cooperate. We see this in some of our *Megachile* and species of other genera.

Bumble bees are considered primitively eusocial. Bumble bee queens emerge from hibernating during winter. The queen finds a new nest hole, often an old rodent hole, and lays her eggs. She then forages for nectar and pollen to provide food for the larvae once they hatch. She does all the work on the colony until those first larvae, her daughters, become adults. Then, the queen devotes herself to egg laying, and the daughters take over the work.

Scientists think that the evolution of sociality in bees may have been made easier by their unusual sex determination system. Unlike humans, bees do not have two sex chromosomes that determine gender. Female bees develop from fertilized eggs, so each female has two sets of chromosomes and is diploid. Male bees develop from unfertilized eggs and have only a single set of chromosomes. This type of sex determination system is called haplodiploidy. After a queen bee has mated, she is able to determine whether or not to fertilize an egg using the sperm she has stored. In social colonies, both worker bees and queens are females. The diet that the bee is fed as a larva determines whether she becomes a queen or a worker. The implication of this for the development of sociality is that workers are more closely related to their sisters than they would be to their offspring so, because workers share more genes with their sisters who may become queens, more of the workers' genes may get passed on to the next generation than they would if the workers reproduced.

Parasites and Robbers

Not all bees forage for themselves. There are bees that actually rob the food stores of other bees. Within the bees, there are three types of bees that behave "badly." First, there are bees that will enter a nest, fight with the owner, and, if they win, take over the nest. This was found in one of the invasive bees in California, *Megachile apicalis*. Some of the social bees actually post guards at the entrances to their nests. These guards probably defend the nest from intruders from the same species intent on taking over the nest, as well as from other predators and parasites. Second, among the social bees there are bees that are called social parasites. These social parasites enter a nest, replace the queen (often by killing her), and then use the workers in that nest to rear their own offspring rather than the offspring originally in the nest. This is very common in the bumble bees *(Bombus)*. Bumble bees have a whole subgenus called *Psithyrus* that are all social parasites. These parasitic bumble bees have even lost their corbicula for carrying pollen. Interestingly, the host bumble bee queen is not always killed when a parasitic bumble bee takes over. She sometimes remains alive. The third and most common type of food-robbing behavior is called cleptoparasitism. Cleptoparasites enter the nests of other bees and lay eggs in the cells of the host bees and then depart. Some of these parasitic adult bees destroy the eggs of the hosts before leaving the cell, but many do not. In some cases, the parasitic egg is hidden in the cell wall of an unfinished cell, but in other cases, the parasitic queen uses her ovipositor to insert her egg in an already closed cell. When the parasitic egg hatches, the larva feeds on the food that had been stored for the host bee. Many of these parasitic larvae hatch with sharp mandibles that they use to kill their host eggs or larvae.

Parasitic bees generally have reduced pollen-carrying structures (scopae, corbiculae) and often resemble wasps.

TABLE 1 Parasitic Bees and Their Hosts

Parasitic genus	Host bee genus
Coelioxys	*Megachile*
Dioxys	*Anthidium, Megachile, Osmia*
Epeoloides	*Macropis*
Epeolus	*Colletes*
Ericrocis	*Anthophora*
Holcopasites	*Pseudopanurgus*
Melecta	*Anthophora*
Neopasites	*Dufourea*
Nomada	*Andrena, Nomia*
Oreopasites	*Calliopsis*
Sphecodes	*Halictus*
Stelis	*Heriades, Hoplitis*
Townsendiella	*Conanthalictus*
Triepeolus	*Melissodes*
Xeromelecta	*Anthophora*
Zacosmia	*Anthophora*

This is because they do not need to have as many hairs for carrying pollen. They also often develop strong cuticles, spines, and stings that are used for defending themselves. Table 1 lists a number of parasitic bees and their hosts.

Other Parasites and Predators

There are numerous nonbee predators and parasites that attack solitary bees. For example, here in California, a study of *Anthophora busleyi,* a bee common to central California, found 22 different organisms in the nests of the bees. Eighteen of those visitors were known to kill bee larvae. Some of the other invertebrates that attack native bees include bee flies, velvet ants, wasps in the families Chrysididae and Ichneumonidae, and oil beetles. Oil beetles and bee flies lay their eggs near bee nest sites and the flowers that bees forage on, and when their larvae develop, they find their way to a bee nest or to a flower. When a bee visits the flower, the quick larvae jump on to her and catch a ride back to the nest. Once at the nest, the larvae eat some of the bee's eggs as well as the nectar and pollen stores. Bee wolves *(Philanthus)* are wasps of the family Sphecidae that prey almost exclusively on bees. The bee wolf carries the prey back to a tunnel but usually stores it only temporarily until it is later used to provision a cell burrow, where an egg is laid. A single bee wolf cell can contain multiple bees. In Massachusetts, a single species of bee wolf, *Philanthus sanbornii,* was documented attacking over 100 different species of bees and wasps.

Some of you have probably noticed small spiders sitting in flowers. These crab spiders ambush bees (Figure 7). The crab spider *(Misumenia vatia),* which is known to catch bumble bees, can change its coloration to match a range of yellow to white backgrounds, effectively camouflaging itself. These are not web-spinning spiders; they are sit and wait predators.

Bees also battle a variety of other invertebrates. These include wax moths that feed on food stores in bumble bee and honey bee nests, flies that consume or parasitize larvae, mites that infest the trachea and feed on hemolymph, parasitic wasps, and nematodes.

There are also larger animals that eat bees. As you might guess, given Pooh's fondness for honey, black bears are well known for raiding beehives. A bear will eat either honey or brood. Other mammals, including badgers, foxes, raccoons, opossums, weasels, house mice, skunks, meadow voles, and other rodents, also attack beehives. They eat pollen, honey, and the bees.

There are also many birds that consume bees. For example, the European Bee-eater, *Merops apiaster,* a gorgeous species found in southern Europe, northern Africa, and western Asia, is a voracious consumer of bees. These birds will catch a bee on the wing and bang the insect on a

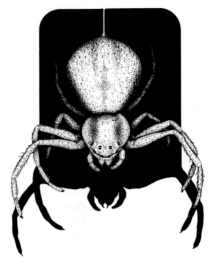

Figure 7. Crab spider

hard surface to remove the sting. One pair of European Bee-eaters can consume 30,000 bees in a season! Bees are often eaten by woodpeckers, tanagers, kingbirds, jays, and other native species here in California.

Nesting

The nests of the bees are where they lay their eggs and their offspring develop. For most California bees, the nest is created by a mother. However, for a few of our social bees, like the bumble bees, workers can help with nest construction. Native California bees are primarily solitary, which means that a single female constructs a nest and lays eggs in it. There are some bees that nest in aggregations where you can find tens or even hundreds of individual nests near each other. Every summer, I get calls about *Anthophora bomboides*, a solitary bee that nests near the beach. Beach-combers are very surprised when they suddenly come upon a nest aggregation and see a lot of good-size bees emerging from the sand and creating small turrets of sand where they have excavated nests.

The basic unit of a nest is a cell or brood chamber. The nest cell protects the developing bee and its food stores, usually a ball of pollen. This nest cell will serve as a nursery for the bee as it transitions from egg to larva to pupa and for its final transition to adult. These cells are often lined. The lining can vary from a cellophane-like substance, to cell walls made of secreted materials that impregnate the matrix, to small pieces cut from leaves. Most nests have more than one cell.

Bees nest in a range of materials, from cement walls to trees to cavities in hollow stems. There are even bees that nest in snail shells. We divide cavity-nesting bees into two main groups; miners or ground nesters, and masons or cavity nesters.

Male bees usually do not sleep in the nest. They sleep outside in places like flowers, tree branches, and grass stems. You can sometimes find clusters of hundreds of male bees sleeping together on a tree branch.

Miners

Bees that nest in mines usually choose a sunny spot with fine soil. The female digs a long shaft that can vary from inches to several feet long. Some nests have a central shaft with cells opening off of it. Other nests have several lateral branches off a main shaft. At the end of the shaft, the bee creates a chamber that will house her pollen ball and egg. She smoothes the surface of the cell or brood chamber using the pygidial plate at the tip of her abdomen and sometimes secretes a substance to line the cell. Once the chamber is built, she will gather enough pollen and nectar to provide for her baby. She often rolls the pollen and nectar into a ball and lays her egg on the ball. Once the egg is laid, she seals the chamber. Many bees create multiple chambers in the same nest, creating an elaborate network of brood chambers off the central shaft.

Cavity Nesters

Cavity-nesting bees usually use a hole created by other species such as beetles. They are called secondary cavity nesters because they are the second users of that cavity. Bees nest in everything from hollow stems to beetle holes to snail shells. In these holes, the female bee creates brood chambers filled with pollen balls just as the mining bees do. Brood cells are in a linear array, with the first cell filled being the deepest in the cavity. Interestingly, the bees often do not emerge from back to front. Usually the innermost cells contain eggs that will develop into females and the outermost cells contain males. The females take longer to develop, so the bees in cells closest to the cavity entrance usually emerge first.

The bee family Megachilidae is called the leaf-cutter family. Those of you who grow rose bushes may have wondered who created the perfectly round holes in your rose leaves. The culprit may be a member of the family Megachilidae. This group generally nests in existing aboveground cavities, but a few either dig holes or use existing holes in the ground. They are called leaf cutters because they often use leaves to both line the inner walls of their nest and cap the end of the nest. If you look closely at a fence, you can sometimes detect a small round hole that has a piece of leaf placed on the end; these are the nests of some leaf cutters. These bees also use mud to line their cells and to construct dividers between the brood chambers. These bees get called masons. A queen mason bee, such as *Osmia nemoris*, will find an area with mud. She will dig until she has enough mud to roll into a ball. She will carry the mud back to her nest in her mandibles. She will then build the dividers for each of her cells, using the mud (much like a mason!), and then cap the nest with a mud plug.

Carpenters

In California, we have only one group that carves its own nests in wood, the carpenter bees *(Xylocopa)*. These bees usually start by digging a tunnel in an overhanging branch (or deck). Once they have bored into the wood, they dig a tunnel to the left and right, creating a T-shaped nest. These bees carve out their tunnels by vibrating their bodies as they rasp their mandibles against the wood. These bees do not eat the wood, though they sometimes reuse the wood particles to create dividers between the nest cells.

Pollination Basics

For plants, bees are a welcome partner. Because plants do not move, they rely on wind, water, or animals to transfer their pollen from the male plant to the female plant. This is called pollination.

Bees go to flowers in your garden to find pollen, a primary protein source, and nectar, which is a sweet liquid that provides energy. Animal-pollinated flowers are usually adapted to attract a particular type of visitor. Most animal-pollinated plants rely on bees. However, bats, birds, moths, butterflies, mammals, and even a lizard have been documented as pollinators. To attract pollinators, plants use color, shape, scents, and various rewards such as pollen and nectar. The flower itself is like a big sign advertising to bees that there is pollen or nectar available—though sometimes a flower will cheat and have nothing! Most bee-pollinated flowers are blue, yellow, or white. This is because bees have blue, green, and ultraviolet receptors in their eyes that do not distinguish reds and pinks well. Bee flowers often also have nectar guides. These guides are usually markings on flowers that guide the bee right into where the pollen or nectar is. They are commonly produced from ultraviolet-absorbing pigments called flavonoids so, when you apply ultraviolet light to a flower or look through eyes like those of bees, these nectar guides show up like targets, directing the bees right to the nectar and placing them in the flower in a way to enhance pollination.

Most flowers have pollen. Bees gather pollen to feed their babies, which start as eggs and then grow into larvae. While some adult bees like bumble bees eat pollen, most pollen goes to the larvae. Bees use the nectar for energy. When bees go to a flower in your garden to get nectar or pollen, they usually pick up pollen from the male part of the flower, which is called an anther. When they travel to the next flower looking for food, they move some of that pollen to the female part of the next plant, which is called a stigma. Most flowers need pollen to make seeds and fruits.

After landing on the flower, the bee moves around. As it moves, some of the pollen on its body gets transferred to the flower. Female and hermaphroditic flowers have a structure called a stigma that serves as a sticky landing pad for pollen (Figure 8). When pollen is transferred to the stigma, the pollen tube grows down the stigma and into the ovary where the unfertilized seeds or ovules are found. Inside the ovule, cells from the pollen join up with cells from the ovary and a seed is created. For many of our garden plants, the only way to start a new plant is by growing from a seed. Fruits are the parts of the plants that have the seeds. While we use the term *fruit* to define a specific group of things in the grocery store, many vegetables such as tomatoes, cucumbers, and peppers are also fruits.

These dustlike pollen grains, the male gametophytes, are produced in an anther, the swollen tip of a stalk-like structure called a stamen. If you look at a lily, the yellow dust that falls from it is made up of pollen grains, and the long, thin stalks with the pollen on the tops are the stamens. For a flower to be fertilized and seed to be produced, pollen (which contains

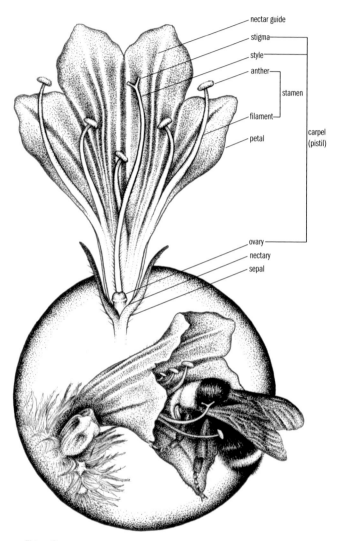

nectar guide
stigma
style
anther
stamen
filament
petal
carpel (pistil)
ovary
nectary
sepal

Figure 8

the male gametes) must be transferred from an anther to the structure that contains the female gametes, called the carpel. Pollen lands on the sticky tip of the carpel, the stigma, which is connected to the ovary by a stalk-like extension called the style. Seeds are produced in the ovary, found in the base of the carpel. A simple flower may have only one carpel; more-complex flowers may have many carpels. Collectively, the female structure of a flower is known as the pistil, regardless of whether there is only one carpel or several.

The androecium, or male part of the plant, is made up of the stamens. Different species of plants have different numbers of stamens, ranging from zero to dozens. Stamens are put together in different ways. Some resemble a ring of fringe surrounding the pistil. This you can see in most open cup-shaped flowers like poppies. Other plants have their stamens fused together. You can see this in a typical tomato flower. Each stamen contains pollen-holding chambers that are fused into an anther. The anther is usually supported on a stalk-like structure called the filament. Each of the chambers in the anther produce the microspores that become pollen.

Adaptations of Bees to Flowers

While there are several other types of pollinators such as bats, birds, and beetles, bees are truly professionals. They have a stunning number of adaptations that really make them good at transferring pollen. Probably the most important trait is their branched hairs. Unlike wasps, which have unbranched hairs (like you might see on a human), bees have hairs that are barbed or plumose, resembling feathers or split ends (see Figure 5).

Bees have specialized brushes or tufts of hairs called scopae that are thought to specifically be designed to transport pollen back to the nest. Different bees carry pollen in different places. The most common body parts for carrying pollen are the hind legs. Some bees have a corbicula—a concave, smooth area on the hind tibia or thorax that is surrounded by long hairs—used to collect and hold pollen that is carried to the nest. This is sometimes called the pollen basket. The term *scopa* usually refers to a brush of hairs used to collect pollen. When bees carry pollen on their legs, they can either carry it dry or mix it with a little nectar. When you look at a bee, it is usually easy to see whether it is carrying dry or wet pollen. Bees in the family Megachilidae carry their pollen on scopae on the underside of their abdomen. You can sometimes be fooled into thinking that a bee has a golden-yellow or even purple underside to its abdomen when, in fact, that color is the color of the pollen the bee has been collecting. When a bumble bee's pollen basket is full, it can contain as many as one million pollen grains. Bees in the genus *Hylaeus* do not carry their pollen externally; they carry pollen internally in their gut.

Bees have other adaptations that hold pollen to their bodies, in addition to hair. A second factor, an electrostatic charge, enables the pollen to stay attached to the insect. The electrostatic charge helps pollen stay on a bee, and recent research has shown that the accumulated charges on bees are sufficient for pollen to be transferred to a flower by electrostatic forces. As a pollen-covered bee enters a flower, the pollen is preferentially attracted to the stigma because it is better grounded than the other parts of the flower. Interestingly, even if the bee does not actually touch the stigma, pollen grains can jump from the bee to the stigma.

Certain bees are able to vibrate or buzz their bodies. For example, bumble bees are very good buzzers, while honey bees and members of the family Megachilidae cannot do this. When bumble bees buzz, they vibrate at a frequency that causes the pores in the anthers of certain plants to open, causing pollen to explode out of the pores, covering the bee with pollen. You can mimic the buzzing of a bee by using a tuning fork. An A440 or middle C tuning fork will work.

Many members of the family Solanaceae, which includes

tomatoes, require buzz pollination. Because bumble bees are semisocial and can have a number of worker bees in a single nest and are able to pollinate tomatoes, a commercial industry has developed supplying bumble bees to tomato growers for pollinating hothouse and—in some states (but not California)—field tomatoes.

Adaptations of Flowers to Bees

If you have ever marveled at the diversity of flower shapes, colors, sizes, and scents, much of that variation can be attributed to the plant's need to transfer pollen. Individual flowers can have both male and female parts, in which case they are called hermaphroditic or perfect, or they can have only male or female parts. Individual plants can have only hermaphroditic flowers, or a combination of male and female flowers, or a combination of hermaphroditic and male or female flowers, or only male or only female flowers.

So why do bees visit plants? In exchange for having their pollen transferred to a new flower, flowers have evolved many rewards to attract pollinators. The most common attraction is food: either nectar, which is a sugar solution, or pollen, which is high in protein. In many plants, nectar is produced in special glands called nectaries. Nectaries are commonly found in the flowers but can be located on leaves or other parts of the plant (extrafloral nectaries). Regardless of where nectar is produced, it is usually protected from evaporation into the atmosphere and from dilution by rainwater. The nectar is also protected from "robbing" by other animals that do not aid in pollination. The concentration of sugar in nectar has evolved to match the energy requirements of the specific animal pollinator. Nectar for bees is typically 10–30 percent sugar, which is similar though slightly lower in sugar content to that produced by hummingbird-pollinated plants and quite a bit lower than that in bat-pollinated flowers. Bees need dilute nectar because their mouthparts cannot suck up thick, syrupy nectar. Sugars are costly to produce, so plants tend to mete out nectar. Some flowers refill when drained; others refill on a daily basis.

Flowers vary in the volume of nectar they produce. Some produce as little as 0.001 ml. If you imagined a worker bumble bee that had a honey stomach volume of 0.1 ml visiting a plant that produced only 0.001 ml of nectar in each flower, you would see that the number of flower visits required to bring back nectar to make a teaspoon (5 ml) of honey is staggering. If half the flowers that she visits have nectar, she would have to visit 200 flowers to fill her honey stomach. Since honey is about half nectar and half water, if her nest is a mile away, she would have to make 100 foraging trips, traveling 200 miles and sucking 20,000 flowers to create a teaspoon of honey! You can see why bees might prefer flowers with higher volumes of nectar.

Plants encourage bees to visit by providing a wide variety of rewards. Pollen, which produces the male gametes of flowers, is also the sole source of protein for bees and is required for larvae to develop. While bees are interested in gathering pollen for their larvae, plants are interested in having their pollen go to the next plant rather than back to the bee nest. Therefore, plants often mete out their pollen in different ways. Some ripen only a part of the pollen in a flower each day; some open only a certain number of flowers each day. Others have elaborate morphologies that allow only certain bees to access the pollen. In addition to nectar and pollen, some of the more unusual rewards provided by flowers to bees are oil, scent, and warmth.

Plants can also be deceptive. Flowers without nectar or pollen often have structures that look just like anthers or mimic a very rewarding flower. Some flowers are shaped like female bees and trick naive males into mating with the flower, although none of these are found in California.

Plants get pollinators' attention primarily through colors and aromas. You can think of many parts of a flower as mechanisms for advertising nectar and pollen rewards. The colors deployed by plants vary depending on the pollinators. Different animals perceive light differently. If you recall, the visible spectrum ranges from 380 to 760 nm (violet–blue–green–yellow–orange–red). Bees see best at the lower end of the visible spectrum and into ultraviolet radiation. This means that a bee's visual acuity is much greater for blues, yellows, and whites than for reds and pinks. Therefore, flowers pollinated by bees are most often violet, blue, or yellow and may have ultraviolet markings (invisible to the human eye), whereas birds are particularly attracted to red flowers and bats' flowers are a dusky white.

Plants also use visual guides (see Figure 8) to direct pollinators to rewards. These are called nectar guides or honey guides. Once the pollinator comes closer to the flower, it sees a visual guide that pinpoints the location of the reward. Some examples of this are the dots on the throat of a foxglove *(Digitalis)* or the lines on the petals of a violet *(Viola)* or a contrasting color pattern like the yellow center and blue petals of baby blue eyes *(Nemophila menziesii)*.

Plants also use aromas to lure in pollinators. Aroma is more important for insect-pollinated than bird-pollinated plants, as most birds do not have a strong sense of smell. Bee mouthparts are covered in tiny hairs that have pores in them. Molecules pass through these pores and stick to receptor sites on sensory cells inside the exoskeleton. This is how the bee tastes and smells. There are also similar hairs on the antennae that pass scent molecules to sensory cells. With two antennae, the bees can monitor the concentration of the odor and locate the source.

A number of compounds have been identified as being associated with bee pollination in particular. Sometimes scent can combine with morphology. For example, the nectar guides of daffodils have a fragrance that is stronger than that of the rest of the flower.

TOXIC FLOWERS

Bees are affected by some other compounds that they find in flowers. Bees are sensitive to ethanol, which causes them to lose their balance and have trouble moving around. It has been reported that very inebriated bees will lie on their backs and wiggle their legs. Bees are also sensitive to alkaloids, coumarins, and saponins, and some compounds like cardiac glycosides are toxic to bees at some levels. Several California native plants are toxic to bees, such as death camas (*Zigadenus*). The California buckeye (*Aesculus californica*) produces pollen and nectar that are reported to be toxic to Western or European Honey Bees but not to native bees.

Aromas do not always have to be pleasant. Some plants actually produce scents that smell like rotting flesh. These odors tend to be more associated with fly and beetle pollination than bees, fortunately.

Bee flowers come in a variety of shapes. An open cup-shaped flower is the easiest for most bees to access. More complicated shapes, like that of a lupine or orchid, serve to limit which bees can access the nectar or pollen in that plant. It takes a strong or heavy bee to open a lupine flower.

Ecosystem Services and Bees

As you sit at the table today, do you know where the water you are drinking came from? In San Francisco, 85 percent of the drinking water comes from the Sierra Nevada. How about the last prescription medicine you took? It probably originated from a natural source. Of the top 150 prescription drugs used in the United States, 118 originate from natural sources: 74 percent from plants that may depend on pollinators, 18 percent from fungi, 5 percent from bacteria, and 3 percent from a species of snake! And where did the ingredients for your lunch and dinner come from? One of every three bites you took probably came from a plant pollinated by wild pollinators, primarily bees. This is just the beginning of a list of the services provided by healthy natural ecosystems.

Economists and ecologists have started working together to find a way to place a financial value on the contribution of natural ecosystems to human existence. The estimates are eye-opening. For example, the value of pollination services to agriculture from wild pollinators in the United States alone is estimated at $4 to $6 billion per year. While these ecosystem services are currently produced for "free," replacing the natural ecosystem would cost many trillions of dollars. Unless human activities are carefully planned and managed, valuable ecosystems will continue to be impaired or destroyed.

Bees and Agriculture

About one-third of the food supply in the United States depends on animal pollinators. While other species like hummingbirds and beetles are involved in pollination, bees are the most important pollinators. Western Honey Bees *(Apis mellifera)* are used extensively as managed pollinators, and native species provide significant pollinator service also. Actually, Honey Bees are not always the best pollinator of a particular crop. Tomatoes are a good example of this. Tomatoes require buzz pollination to release their pollen. Since Honey Bees do not buzz, they are not good pollinators for tomatoes. Table 2 lists some bee-dependent crop plants in California and the bees that pollinate them.

If you are going to talk about pollination and agriculture in California, it is best to start with almonds. Approximately 80 percent of the almonds produced in the world are grown in California. The 2009 estimate of the value of the

TABLE 2 Some Bee-dependent Crop Plants in California

What crops do bees pollinate?	Which bees?	What is produced?	How necessary are bees?
Almonds	Honey bees, bumble bees, solitary bees (Osmia)	Nuts	Essential
Apples	Honey bees, bumble bees, solitary bees (Andrena, Halictus, Osmia, Anthophora)	Fruit	Great
Avocados	Honey bees, solitary bees	Fruit	Great
Blueberries	Honey bees, bumble bees, solitary bees (Anthophora, Colletes, Osmia ribifloris, Osmia lignaria)	Fruit	Great
Canola	Honey bees, solitary bees (Andrena ilerda, Osmia cornifrons, Osmia lignaria, Halictus)	Seeds	Helpful
Cherries	Honey bees, bumble bees, solitary bees	Fruit	Great
Prunes	Honey bees, bumble bees, solitary bees	Fruit	Great
Pumpkins	Honey bees, squash bees, bumble bees, solitary bees	Fruit	Essential
Soybeans	Honey bees, bumble bees, solitary bees	Seeds	Helpful
Squash	Honey bees, squash bees, bumble bees, solitary bees	Fruit	Essential
Strawberries	Honey bees, bumble bees, solitary bees (Halictus)	Fruit	Helpful
Sunflowers	Honey bees, bumble bees, solitary bees	Seeds	Modestly helpful
Tomatoes	Bumble bees, solitary bees (Halictus)	Fruit	More or less helpful, depending on variety
Watermelons	Honey bees, bumble bees, solitary bees (Peponapis)	Fruit	Essential

almond crop to California's economy was $1.9 billion. That is a lot of almonds! Almonds are primarily pollinated by Honey Bees. They bloom in February, which is early enough that the majority of native bees have not emerged. However, when there are native bees in the almond orchard, Honey Bees change their behavior and become much more effective pollinators because they are more likely to move among orchard rows.

Each winter, virtually every commercial Honey Bee hive west of the Mississippi River (about one million Honey Bee hives) is shipped to California. Hives are then placed out in the almond orchards waiting for the flowers to open. Farmers put out about two hives per acre, and the cost of Honey Bee hive rental in 2011 was about $50 per hive for almond pollination. Almonds are a major source of income for both the almond farmer and the beekeeper.

Most of the bees that are managed for agriculture are cavity-nesting bees like Honey Bees and leaf-cutter bees. This in part is because they can be moved around by farmers who need their services. The only managed ground-nesting bee is the Alkali Bee, *Nomia melanderi*. This bee has been used to pollinate alfalfa *(Medicago sativa)* for over 50 years. In some areas where alfalfa is grown for seed, notably the Touchet Valley in Washington State, which grows about 25 percent of the US seed, soil-nesting beds for *N. melanderi* are actively managed. Growers subirrigate and surface-salt the areas where the bees nest. And, it is wildly successful. On a single 3-acre (1.5-hectare) nesting bed, there is an aggregation of over five million bee nests with up to 1,000 nests in a square meter. This is the largest bee nesting aggregation ever recorded.

Bees in Urban Environments

To maintain biodiversity and to meet the increasing demands for ecosystem services, we must move conservation science into cities. Cities are important for conservation for two reasons. First, 80 percent of the US population already lives in urban areas. Second, cities are growing. They already encompass about 3 percent of land (59.6 million acres) in the United States, and 230,000 additional acres become urban each year. Because of their large human populations, cities are the places where many ecosystem services, such as environmental quality of life, are delivered. Given the growth of the urban population, it is clear that we need to develop the knowledge necessary for maintaining natural habitats in the urban setting and find a way to give urban dwellers access to nature.

We know that pollinators are declining in certain wild and many agricultural landscapes. However, little is known about urban pollinators. Our recent data on bumble bees in San Francisco suggest that urban bees may also be declining. While the loss of these pollinators is important, it is also important to understand what effect these losses have had on pollinator services. A recent study on cherry tomatoes in San Francisco suggests that without pollinators, there would be lower yield from each plant. In other words, the plants' tomato production seems to be determined by the supply of pollen.

We do not know much about how healthy bee populations are maintained in an urban environment. Because natural habitats are less common in urban landscapes, they may not provide enough resources to support viable pollinator communities. However, if other habitats, such as urban gardens and restored areas, are sufficiently connected to natural habitat, then native populations may thrive. Work in seven California cities (Ukiah, Sacramento, Berkeley, Santa Cruz, San Luis Obispo, Santa Barbara, and La Cañada Flintridge) by Gordon Frankie suggests that in California, urban environments can provide habitat for bees. He found that urban environments support approximately 60–80 species from all five families found in the state. In these settings, the bees often used nonnative ornamental plants as their food sources. While 60–80 species is much less than the 100–300 you might expect to find in the surrounding rural habitats, they are still good pollinator communities.

Conservation of Bees and Other Pollinators

Bees have complex life cycles and specific habitat needs. As such, bee populations are threatened by habitat destruction and fragmentation, pesticides, climate change, and a host of incompatible land management practices. Here are some things you can do to help bees.

PLANT A POLLINATOR GARDEN One of the best things to do for bees is to plant a pollinator garden. The Xerces Society, an invertebrate conservation organization, has developed guidelines for improving habitat in situations ranging from a backyard to a golf course to a farm. The guidelines here are adapted from their recommendations.

Flowers are food resources. Patches of habitat can be created in many different locations, such as field edges, stream corridors, backyards, school grounds, golf courses, street medians, and city parks. Even a small area planted with the right flowers will be beneficial because each patch will add to the mosaic of habitat available to bees.

INCORPORATE A SUCCESSION OF FLOWERS Doing this will provide bloom throughout the entire growing season. A diverse selection of flowering plants that bloom in succession will support a more diverse community of pollinators.

CHOOSE SPECIES THAT ARE EASY TO PLANT AND ESTABLISH Your local chapter of the California Native Plant Society and native plant nurseries can give advice on local plant species. For California, there is a list of urban bee plants.

USE LOCAL NATIVE PLANT SPECIES Native plants are more likely to attract native bee pollinators than are exotic

TABLE 3 Excellent Plants for Native Bees

Common name	Scientific name (genus)	Common name	Scientific name (genus)
Aster	*Aster*	Penstemon	*Penstemon*
Black-eyed susan	*Rudbeckia*	Purple coneflower	*Echinacea*
California poppy	*Eschscholzia*	Rabbitbrush	*Chrysothamnus*
California redbud	*Cercis*	Rhododendron	*Rhododendron*
Chamise	*Adenostemma*	Sage	*Salvia*
Creosote bush	*Larrea*	Scorpion weed	*Phacelia*
Currant	*Ribes*	Snowberry	*Symphoricarpos*
Elder	*Sambucus*	Sunflower	*Helianthus*
Goldenrod	*Solidago*	Toyon	*Heteromeles*
Huckleberry	*Vaccinium*	Wild buckwheat	*Eriogonum*
Lupine	*Lupinus*	Wild lilac	*Ceanothus*
Manzanita	*Arctostaphylos*	Willow	*Salix*
Oregon grape	*Mahonia*		

flowers. You may supplement natives with heirloom varieties of herbs and perennials. Do not plant invasive species.

CHOOSE FLOWERS WITH A VARIETY OF SHAPES AND COLORS This will attract a wide variety of bees. Different bee species have mouthparts adapted to different shapes of flowers; short-tongued bees can drink only from open flowers such as asters or daisies, while long-tongued bees can reach the high-energy nectar in deep flowers such as bluebells or lupines. Flower colors known to attract bees are blue, purple, violet, white, and yellow. We recommend that you use native plants wherever possible. We have listed some of the native plant genera most attractive to bees in Table 3.

PLANT FLOWERS IN CLUMPS Flowers clustered into clumps of one species will attract more pollinators than individual plants scattered through the habitat. Where space allows, make the clumps 4 ft or more in diameter.

Land Management Considerations

You can help maintain healthy populations of bees by making slight adjustments in land management practices to ensure habitat is provided throughout the year. In order to maximize pollination opportunities, avoid maintenance treatments while plants are in flower, and allow plants to bolt before tilling. Never burn, graze, or mow an entire area at once. This will allow for recolonization of the treated area from unaffected areas, an important factor in the recovery of pollinator populations.

AVOID PESTICIDES Do not use pesticides unless absolutely necessary; insecticides can kill bees, and herbicides can kill their food sources. If they must be used, pesticides should be applied with careful timing and targeted spraying methods in order to minimize impacts on pollinators.

PROTECT NEST SITES Solitary ground-nesting bees need direct access to the soil surface to excavate and access their nests. Where possible, keep some bare or partially vegetated ground. Clear the vegetation from small patches of level or sloping ground, and gently compact the soil surface. These patches should be well drained and in open, sunny places. Ground-nesting bees seldom nest in rich soils, so poor-quality sandy or loamy soils may provide fine sites.

Do not put thick layers of mulch or landscape fabric over potential nests sites. These barriers can prevent bees from burrowing into the soil.

Create sand pits and piles. In a sunny, well-drained spot, fill a pit about 2 ft deep with a mixture of fine-grained sand and loam. Where soils do not drain well, make a pile of the sand-loam mixture instead—or fill a planter with the mixture.

Other solitary bees nest in solitary wood (or tunnel). For them, leave snags or dead tree limbs. An arborist can advise on whether a snag is really a hazard; if you can leave it, it

About a third of our bees, such as the family Mega-chilidae, nest in small cavities, often the tunnels of beetles. For these bees, leaving tree snags in your yard is a great way to keep nest sites available. If you cannot keep snags, one alternative is to build nest blocks (or buy them at a local garden center). Nest blocks can easily be made from old lumber. You will want to use untreated lumber and drill nest holes for the bees to use. Different species of bees nest in different-size holes, so I recommend you use different sizes of drill bits ($\frac{3}{32}$ to $\frac{3}{8}$ in.). Drill the holes 5 or 6 in. deep. Each hole should be at least half an inch from the next. The inside of the hole should be smooth and closed at one end. You don't have to use lumber for this; you can use logs or stumps. When you have the bee condo made, you can mount it. It should be somewhat sheltered and face a direction that will give it morning sun. Fence posts, trees, and buildings can all be used to mount your condo.

There is a variety of designs for bumble bee nest boxes, as well as commercially available premade boxes. These boxes have very low rates of success. A better way to attract bumble bees would be to make improvements to your garden to enhance bumble bee habitat.

---

may provide excellent nest habitat. Cut the ends off pithy-centered stems on plants such as elderberry, box elder, raspberry, or dogwood. Bees will nest in holes they dig in the soft pith. Supplement your garden with artificial nest boxes (such as a nest condo).

Studying Bees

Studying bees through either observing or collecting them will provide wonderful insights into this marvelous group.

Fortunately, it is easy to study bees in your own backyard. Bees are most active in spring and summer months when floral resources are available. Bees can be found everywhere in California from high mountain meadows to wind-swept dunes. The easiest places to find bees are the places where there are flowers. However, if you watch only flowers, you may miss some of the more interesting parasitic species that are more often found patrolling bare ground, looking for the nests of their hosts. Bees prefer open habitats. If you are new to an area, then looking along roadsides, river edges, power lines, open fields, sandy areas, and wetlands will be a great way to start.

Watching Bees

You can start by simply watching. Some of the binoculars that can focus at close range are fantastic for watching bees. One of the first things to notice is what the bee is doing on the flower. Some bees will simply be taking a sip of nectar, while others will busily be collecting pollen. You can watch and see how they move their legs, how quickly they depart after landing, how they move the pollen around their bodies. As you watch different bees on different shapes of flowers, you will begin to notice where on its body a bee is placing the pollen it collects. Sometimes you will see a bee with pollen on its face. Others will have it sprinkled on the underside of the abdomen or on its hind legs. Pollen grains that are not gathered up into pollen baskets are the key to the plants' reproductive success, for these are the grains that get transferred to the next flower.

If you are lucky enough to find a nest, you can watch, or time, how long it takes for bees to return. Are their foraging trips long or short?

You can also participate in more formal activities like the Great Sunflower Project (www.greatsunflower.org). By using citizen scientists across the continent, all surveying pollinators in backyards, parks, and gardens, the Great

Sunflower Project gathers the necessary data to address questions about the effect that declines in bee populations have had on pollination of plants. In addition, by examining pollination in their own gardens, participants get the ability to place their gardens in local, regional, and even continental contexts. Imagine being able to know enough about pollinator service to say, "Wow, my garden has many more pollinators than most of my neighbors and most of San Francisco, but fewer than average for the Bay Area in general, California, and the western United States."

The Great Sunflower Project uses sunflowers as standardized "bee-o-meters"; for each site with sunflowers, citizen scientists record how many bees visit within the time that they watch, effectively creating an index of pollinator service. The Project also accepts data from other flowers and even counts of bees taken along trails or walkways. By also taking and uploading photographs of the visiting bees, citizen scientists help collect the data that will answer whether or not native bees are filling in where we have had a loss of Western Honey Bees through colony collapse disorder or other maladies.

Collecting Bees

Insect collections are incredibly valuable resources for science. Not only do they provide documentation of the presence of species, they can be used to understand the relationships among species, how organisms evolve, and contribute to many other fields. For many insects, bees included, accurate identification of the species requires using a microscope. Because of this, we do need to collect bees. However, collections should be done thoughtfully. While there are some wonderful experiences to be had building personal collections, collected insects are most valuable in collections where scientists can get access to them. If you are interested in collecting, we would suggest working with local professional or experienced amateur entomologists and, if possible, making arrangements to donate your specimens to a local museum to ensure that the bees that you kill are able to make a contribution and are available for all to see and use.

To make a bee collection, you will need some equipment. The first item is an aerial net that can be used to catch flying bees. Most bee collectors use nets with a 12–18 in. hoop and a 3-foot handle. The larger hoop has an advantage because it covers more area; however, it is harder to swing and more likely to get snagged. I actually use a net made from an old golf club shaft because the aerodynamics are great. You can also find collapsible nets commercially. Some break down into three pieces; others telescope. The net should have a net bag that is made out of canvas material around the rim and wedding veil for the body of the net. You can either make this yourself or buy a commercially available one. Since many interesting bees are tiny, it is useful to buy or make a net bag that uses a very fine mesh.

NETTING A BEE It is best to carry your net so that you are ready to swing. I hold the handle of the net in one hand and the middle of the shaft in the other. I also usually hold the tip of the net bag in the hand on the shaft. When looking for a bee to catch, it is easier to detect the movement of a bee than the bee itself. Swing at everything. There is no harm in catching insects that are not bees, so I worry less about identifying what is moving than catching whatever it is. It is important when you swing your net that you swing quickly and swing all the way through—much like you would a tennis racket or a golf club. The only time that I check my swing is when I am working around plants with thorns or seeds that might damage my net bag. As I walk, I try to pay attention to the direction the wind is blowing. Swinging into the wind will open your net bag up, making it more difficult for bees to escape. In some instances, it is going to be easier to capture a bee by slapping your net on the ground than swinging through. This is especially true when you have small bees working plants that are near the ground. In this situation, I slap the net on the ground and then lift the end of the net bag in the air. Usually, any bee in the hoop will fly up the net bag trying to escape.

When looking for bees in a patch of flowers, I am usually conscious of where my shadow is. Bees will move away from shadows. It is best to stand several feet away from the flowers but close enough that you can quickly swing at a bee that comes into sight without taking a step. I often watch to see the pattern of flight in a patch. You will see some bees quickly flying through the patch without stopping. These are often male or parasitic bees looking for females. To catch these bees, you need to be quick. You often have a second chance though, as they seem to repeat their circuit with some regularity.

GETTING A BEE OUT OF THE NET Once you have actually captured a bee, you need to find a way to get it out of the net. If you are simply going to look at the bee and not collect (a euphemism for kill) the bee, you can use a glass jar to capture the bee. I often use a test tube with a cork inserted in the end, as I find it is easier to slip a cork in than to screw on a top with one hand. If you bring a cooler of ice with you, you can place the jar in it and the bees will become inactive and easy to examine—that is, until they warm up!

If you are going to collect the bee, you will want to have something to kill the bee. Traditionally, collectors have used collecting jars. Collecting jars are usually made with cyanide or ethyl acetate. If you are using a collecting jar, it helps to have a second set of containers to transfer dead bees to every half hour or so. The chemicals in a collecting jar can change the color of a bee and sometimes make the body stiff, rendering it more difficult to pin. I use a set of small plastic containers designed for holding beads that I purchase at a craft store. A simpler and less toxic way to collect bees (though it takes a bit more processing back in the lab) is to use a jar filled with soapy water. One or two drops of dishwashing liquid per cup of water is sufficient.

Once the bees are in the bag, I swing the net rapidly to "snap" the bees to the end of the net bag. I then trap the bees there by grabbing the net below where the bees are. The netting of the bag will bunch up and protect your hand from getting stung. I then slip the collecting jar into the net and capture the bees.

WASHING AND DRYING BEES Bees that are collected in liquids should be washed before being pinned. The easiest way to do this is to get a mason jar and replace the top with window screen. Add enough warm water to cover the bees but do not fill the jar. If your bees were collected in something other than soapy water, you should add a squirt of dishwashing liquid to the container and then shake vigorously for more than a minute. Next, rinse the bees with water until the water runs clear.

Once you have done this, rinse the bees in alcohol and then shake them out onto a paper towel to get the initial water and the alcohol off. Then, move the bees to a stack of dry towels and fold the towels over the bees to rub them dry. You can rub with some vigor without damaging them. Having a stack protects the specimens a bit. Pick up the top towel by its corners, and bounce the bees around in the basket that forms for a minute or more. This step helps to fluff the hair on the bees and really makes your specimens look better. You can also use a blow-dryer or build your own bee dryer to dry your bees, and instructions are easily found on the Internet.

PINNING BEES Collected bees are usually stored on pins. Bees are pinned using entomological pins (sizes 1–3). Size 2 pins are useful for virtually all bees that we have in California. Bees are usually pinned directly from the collecting jar or drying towel into a box. Bees are pinned in the thorax right between the wings. Usually, the pin is inserted through the right side of the bee, leaving the left side intact. The bee is positioned toward the top of the pin, leaving enough room at the top for someone to safely grab the pinned specimen without contacting the bee. You can use a pinning block to determine the correct height.

STORING BEES There is a variety of commercially available boxes, drawers, and storage cabinets that can house your collection. For temporary collections, I use cardboard specimen boxes. These have the advantages of being inexpensive and easy to store, and you can write on them. For more-permanent collections, I use a glass-topped hinged museum drawer.

When pinning into a storage box prior to making permanent labels, I start by creating a label with the date, site, and collection ID. This is often the tag that I slipped into the vial when I was collecting the insects in the field. I then pin all the insects from that site to the right of the tag. I signal the end of that group by placing the next tag next to the final specimen. In general, I try to keep bees from a single collecting trip or project in the same box. On the outside of the box, I write the appropriate identifying information so I can quickly relocate the specimens. You will want to take some steps to keep pests, particularly dermestid beetles (family Dermestidae), out of your collection. You can do this by regularly freezing (for 3 days) your boxes, by keeping a mothball or other chemical deterrent in the box, or by simply storing the box in a plastic ziplock bag. We find that a 2 gal ziplock is the right size for most standard insect boxes. If you live in a very humid area, you will also want to be careful about mold and mildew.

Identifying Bees

It is almost impossible to identify bees while they are flying, so to identify a bee you need to be able to take a good look at it. If you are not planning to collect the bee, you can often get a good look by simply cooling the bee down. Put the bee in the refrigerator or in a cooler with ice, and wait 5 minutes or so. This will immobilize bees, although you need to be aware that they will become active again as they warm up. Then, you can use a hand lens or dissecting microscope to look at the bee. It helps to place your bee on a white piece of paper because this will allow you to look at some of the characteristics of the wings more closely. You can use the matrix in Appendix 2 to differentiate among the genera. The matrix starts with the size of the bee. Is the bee larger or smaller than your average Western Honey Bee worker? Then you will want to examine the color of the bee. Bee colors can be subtle; to distinguish among the dark blue, black, and green colors, it can be easier to look at the sheen of the bee. Next, look for markings on the bee. These are most common on the face or the abdomen of a bee.

HONEY BEES

Arguably the most familiar bee to humans is the honey bee. Honey bees are not native to California. We think honey bees originated in eastern Africa and migrated from there into Europe and Asia. The species we see in California is sometimes called the European Honey Bee, but a better name might be the Western or European Honey Bee (see the genus account for *Apis*).

Some of the earliest records of the relationship between bees and humans come from late Paleolithic cave paintings like those of the Cuevas de la Araña found near Valencia, Spain. Ancient beehives have also been discovered in Israel that date back 3,000 years, and Egyptian beekeeping operations are depicted on ancient tomb paintings. In North America, the first honey bees were brought to Jamestown, Virginia, by European settlers in 1622. Interestingly, these were not the first members of the genus *Apis* (the honey bees) in North America. A honey bee fossil of a different species *(Apis nearctica),* complete with hairy eyes, was found in Nevada. This species may have crossed to North America from Asia in the Beringia migration about 30,000 years ago and gone extinct due to changes in climate at the end of the Miocene era. *Apis mellifera*, our Western Honey Bee, has spread now to most continents and is the most important partner for agriculture.

Western Honey Bees are considered social bees. This means that there is a social differentiation among the bees based on their functions. The division of labor in a Honey Bee hive is complex and fascinating and worthy of a book unto itself. Every Honey Bee in a hive belongs to one of three castes: queens, female bees called workers, or male bees called drones.

In each hive there is a single queen whose main purpose is to lay eggs. She is able to lay over 1,500 eggs per day. The queen lives for 2 to 8 years. She is larger than the workers and drones. Her ovipositor is smooth and curved, so she can use it over and over again.

The workers are the daughters of the queen. Different workers do different tasks in a hive. When they are young, they work within the hive constructing comb, managing the brood, cleaning, regulating hive temperature, and defending the hive. As the workers age, they become foragers. These foragers gather nectar, pollen, and water from outside the hive. Interestingly, as workers age, certain regions of their brain increase in size. It is thought this may be associated with their spatial learning from foraging. During the main period of hive activity, a worker only lives about 6 weeks. Workers are also morphologically different from the queen. Aside from being smaller, they have a corbicula (pollen basket) on each hind leg, an extra stomach for carrying nectar, and special glands on their underside for secreting beeswax. Their ovipositor also differs, as it is straight and barbed. This is why you always hear that a bee dies after stinging. A worker Honey Bee's stinger is ripped out of her abdomen when she stings, which kills her.

Drones are male bees. They, therefore, have no ovipositor. The only reason for drones to exist is to mate with new queens. Drones have bigger eyes than workers and queens and have longer abdomens. They lack the specialized structures for collecting pollen and nectar.

The first Honey Bees brought to North America were kept in hives and managed for their honey production. However, Honey Bees have escaped from management and successfully colonized the United States. In 1862, L. L. Langstroth, a Philadelphia minister and beekeeper, used the knowledge that Honey Bees will fill in any gaps in their hives that are larger than 1 cm, a concept called bee space, and invented the modern beehive with movable frames. This invention really led to the development of modern apiculture, as it allowed beekeepers to harvest honey, move their hives, and manipulate their colonies. Here in California, we have both escaped wild Honey Bee colonies and managed colonies.

continued →

Wild hives can be found in the walls of houses, in cavities in trees, or in any large cavities. Wild hives build honeycomb and produce and store honey just like managed hives. When you see a swarm of Honey Bees, these are bees that have left their natal hive (either a wild hive or a managed hive) and are following their queen as she searches for a new home. While swarming, the queen is emitting a pheromone that is all but irresistible to the accompanying worker bees, and they will follow her for miles and days.

The swarm will eventually settle into a new home. Sometimes feral swarms are captured by beekeepers and moved into artificial hives; other times, the swarm finds its own cavity. Once the queen settles in, the workers begin to construct honeycomb from wax and to provision that comb with pollen and nectar—for the new eggs that the queen will lay and to provide food resources for the bees of the hive. Honey represents the food storage of the hive. Individual worker bees bring nectar back to the hive. They regurgitate that nectar into a cell in the honeycomb. Once in the honeycomb, the water from the regurgitated nectar evaporates, making it thicker syrup. Bees can accelerate this evaporation by fanning the comb with their wings. Once the honey in a cell reaches the right consistency, the bees plug that cell with wax, and the honey is stored until needed.

In California, wax is only produced by bumble bees and Honey Bees. It is produced from eight internal wax glands found in the abdomen of the bee's body. As bees age and begin to take on the role of forager, the wax glands decrease in size. The wax produced by the bee is clear; after the bee chews it, it becomes white, and then as pollen oils and propolis are added, the wax becomes yellower and browner.

HONEY BEE PROBLEMS

Over the past 60 years, Honey Bee colonies have declined by 59 percent because of the effects of a parasite, the Varroa

Mite, among other things. The USDA and Apiary Inspectors of America have reported that in both 2006 and 2007, commercial beekeepers in the United States lost a little over 30 percent of their Honey Bee colonies. Much of this loss has been attributed to colony collapse disorder, a condition that results in the rapid loss of a colony. As far as we know, colony collapse disorder does not affect other species of bees, only Western Honey Bees.

COLONY COLLAPSE DISORDER Honey Bee hives have been known to collapse at several different points in the history of apiculture. However, the term *colony collapse disorder* was first used for the big decline in Honey Bee colonies in late 2006. Initial reports attributed colony collapses to factors that ranged from cell phone usage to the development of new pesticides, particularly neonicotenoids such as imidacloprid that affect the central nervous system of insects. Surveys looking at differences between colonies affected by colony collapse disorder and ones that are not have shown that the affected colonies have higher levels of viruses, pesticides, and parasites, suggesting that environmental stress may lead to a series of events within the colony that render it more susceptible to parasites and pathogens. More recent genetic work has found that affected colonies all are infected by both a virus (invertebrate iridescent virus type 6) and a spore-forming (microsporidian) fungus *(Nosema ceranae)*.

HONEY BEE ENEMIES In addition to the players implicated in the studies of colony collapse disorder, Honey Bees have a variety of predators, parasites, and diseases. Individual bees are eaten by birds, bats, flies (family Asilidae), dragonflies, wasps, and other organisms. Social bee nests, which have lots of food resources as well as tasty larvae, are raided by everything from ants to raccoons to bears (which my student Amber discovered after placing bumble bee nests in high Sierra meadows).

I've provided two tools to help you identify different genera of bees. Below there is a key of characteristics that can help you sort among genera. In addition, Appendix 2 includes a detailed list of key characteristics. Many of the characteristics require a microscope or a strong hand lens to be seen.

Key of Basic Bee Characteristics

Here are some quick ways to get at some of the most common genera using some simple characteristics:

Two submarginal cells in the fore wing.................
...... see Family Megachilidae, and Genera *Hylaeus, Dufourea, Panurginus,* or *Perdita*

Obvious pollen on the undersurface of the abdomen......
.... see Family Megachilidae

Very large size, black or metallic ..... see Genus *Xylocopa*

Large size, black and yellow (sometimes red/orange)......
.... see Genus *Bombus*

Medium to large size with conspicuously hairy, brush-like hind legs (scopae)... see Genera *Melissodes, Anthophora, Peponapis,* or *Xylocopa*

Medium-size bee with elaborate white or yellow patterns on abdomen................ see Genera *Anthidium, Dianthidium,* or *Triepeolus*

Small black bee with white markings on face.............
.... see Genera *Hylaeus* or *Ceratina;* see also male *Andrena* or Family Halictidae

Bright metallic green or blue ....
.... see Genera *Agapostemon, Osmia,* or *Hoplitis*

Strongly heart-shaped "face" .... see Genus *Colletes*

Wasplike, hairless..... see Genera *Nomada* or *Sphecodes*

Bee found in a squash blossom...... see Genus *Peponapis*

If you are really interested in learning to identify bees, one of the best things to do is to work with already identified specimens. You can often find these at local museums or universities. As you will know what bee you are looking at, you can see what errors you make and learn from them. It is difficult to be certain your identifications are correct without what we call a reference collection.

In the following family and genus accounts, you will find a general summary for each genus, followed by a description that provides details on how to tell the genus from other genera, similar insects that you might mistake for the genus, food resources, information on nests, and timing of flight. Appendix 1 provides a California species list. Appendix 2 provides a key to females of each genus included in the book.

PLASTERER or POLYESTER BEES
(Family Colletidae)

Family Colletidae is a large, mostly Australian family with a worldwide distribution (excluding the Arctic and Antarctic). This was once considered the most primitive family of bees, but recent molecular work suggests that the family Melittidae may be more primitive. Understanding the behavior of these primitive groups is important in understanding how bees coevolved with plants. For example, if the Colletidae are the most primitive, it suggests that early bees were generalist foragers, whereas if Melittidae are the earliest ancestors, early bees were plant specialists, that is, they fed on only specific plants, and later groups of bees evolved to become generalist foragers.

Colletidae come in a variety of shapes and sizes, but most bees in this family have a short, broad, two-lobed glossa and a short, broad labrum. They also have only one suture below their antennal socket. They are often called plasterer bees because they line and smooth the walls of their nest cells with cellophane-like secretions. Many of the bees in this family do not roll balls of pollen. Instead, they create liquid or semiliquid pollen masses on which the larvae develop. Most Colletidae are solitary and very rarely parasitic.

There are seven genera in the family Colletidae in North America: *Caupolicana, Colletes, Crawfordapis, Eulonchopria, Hylaeus, Mydrosoma,* and *Ptiloglossa.* Two are known from California: *Hylaeus* and *Colletes.*

YELLOW-FACED BEES or MASKED BEES

Genus *Hylaeus*

Common California species: *Hylaeus modestus*   `NATIVE`

GENUS SUMMARY: *Hylaeus* (hi-LEE-us) is a genus that is found worldwide, with about 700 identified species. There are approximately 25 species in North America and 14 in California. The individual species tend to be widely distributed and can be very abundant locally.

DESCRIPTION: *Hylaeus* are shiny black, slender, hairless, and superficially wasplike bees. They are small bees, 0.2–0.3 in. (5–7 mm) long. These markings are more pronounced on the males. This group is called masked bees because of the small yellow or white diamond-shaped markings on their face.

Hylaeus do not have colored markings on the abdomen. They lack pollen-carrying scopae and instead carry both pollen and nectar internally. They have two submarginal cells in their fore wing. Under magnification, you can see the basal vein is not strongly arched at the base. The jugal lobe of the hind wing is about three-fourths as long as the vannal lobe. They have narrow slots or foveae running along each compound eye from the base of the antenna to the ocelli. The labrum is broader than it is long. *Hylaeus* females have one subantennal suture.

SIMILAR INSECTS: *Hylaeus* resemble small sphecoid wasps. *Hylaeus,* like all bees, have branched hairs that do not reflect light, while wasps have unbranched, light-reflecting hairs that glitter in light. You can only see these hairs under magnification. *Hylaeus* also resemble *Ceratina* but are not quite so robust or shiny.

FOOD RESOURCES: *Hylaeus* carry pollen and nectar in their gut, not externally, and regurgitate it upon returning to their nest. This way of carrying pollen makes it difficult to assess what flowers they visit (because pollen cannot be sampled from the bee without dissection). They are suspected to be

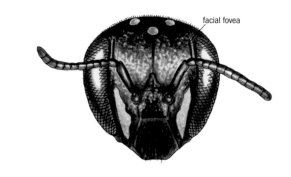

facial fovea

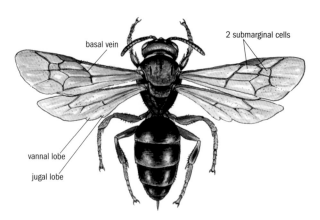

2 submarginal cells

basal vein

vannal lobe

jugal lobe

Hylaeus modestus, native

primarily generalist foragers. *Hylaeus* are short-tongued, but their small body size enables them to access deep flowers. The pollen wasps, members of the genus *Pseudomasaris,* also carry their pollen and nectar internally.

NESTS: *Hylaeus* nest in stems and twigs, lining their brood cells with a self-secreted cellophane-like material. There are a few species that use preformed cavities. They lack strong mandibles and other adaptations for digging; thus, many species rely on nest burrows made by other insects.

FLIGHT SEASON: This genus has primarily spring bees though some persist into fall.

POLYESTER BEES or DIGGER BEES

Genus *Colletes*

Common California species: *Colletes hyalinus*   NATIVE

GENUS SUMMARY: *Colletes* (koe-LEE-teez) is a widespread and common genus with approximately 460 identified species worldwide. There are 98 species in North America and 31 species from California. They range as far north as Alaska.

DESCRIPTION: *Colletes* are small to moderately large bees, 0.3–0.6 in. (7–16 mm) long, with a very hairy head and thorax. They have pale bands of hair on their abdomen. Their face seems to taper toward the mouth, and the eyes are slanted toward each other, making their head appear heart-shaped. They carry pollen in scopae on the upper to lower part of the hind legs. They have three submarginal cells and are the only genus with an S-shaped second recurrent vein on the fore wing, which you can see under magnification.

SIMILAR INSECTS: *Colletes* resemble some *Andrena* and *Halictus,* but the head of *Colletes* is more heart-shaped. *Colletes* have a curvy vein in their fore wing that also helps distinguish these bees.

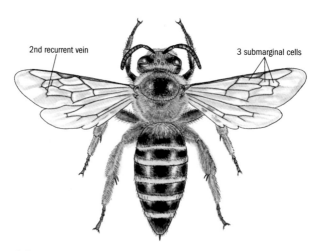

2nd recurrent vein

3 submarginal cells

Colletes hyalinus, native

FOOD RESOURCES: Many *Colletes* are floral specialists and may visit only a small number of plant species. The specialist bees are generally found on plants of the families Asteraceae, Papilionaceae, Hydrophyllaceae, Boraginaceae, Malvaceae, Zygophyllaceae, and Salicaceae in California.

NESTS: Most *Colletes* make solitary nests in the ground, and a few species nest in large aggregations. Often called polyester bees, *Colletes* have a unique method of lining their brood cells with a completely waterproof cellophane-like material secreted from their Dufour's gland. The cellophane-like material does not permeate the surrounding ground; it is instead easily separable from the soil. A *Colletes* female completely encloses her brood cells in this waterproof membrane, thus protecting her brood from fungal attack. *Colletes* also secrete linalool, a fungicide and bactericide used to protect brood cells, from a gland near their mandibles. *Colletes* leave provisions for their young in liquid form (like *Hylaeus* species) and attach the egg to an upper wall of the brood cell rather than placing it on the provisions.

One species of fly, *Miltogramma punctatum,* is a specialist parasite on some *Colletes* species. It follows a female *Colletes* back to her nest and waits near the nest entrance for the bee to leave. The fly then enters the nest and deposits its egg in a *Colletes* cell. The fly egg soon hatches, and the fly larva eats the pollen and nectar in the cell, causing the newly hatched *Colletes* larva to starve.

FLIGHT SEASON: This genus has a mix of spring and fall bees. There is even one species group that produces two broods a year, one in spring and one in fall.

SWEAT BEES (Family Halictidae)

Family Halictidae is found worldwide. Most members of the family are of small to medium size 0.1–0.6 in. (4–14 mm) and are generally dark colored, though a few are bright green and some are red. Several species have yellow markings, particularly in the males, who often have yellow facial markings. Members of the family Halictidae are morphologically very similar. They have even been called boring because they look so much alike. Even so, they have an amazing diversity of social behaviors, with some species having queen and worker castes (though not as complex as those in the Western Honey Bees). Some species are social under certain conditions and solitary under others. The family is most easily distinguished under magnification. One characteristic that would suggest you are observing a member of this family is the presence of an arch toward the base of the basal vein on the fore wing. Some members of Halictidae forage in the evening and into the night, which is unusual for bees.

There are 34 genera in the family Halictidae in North America: Several genera within this family are commonly observed in California: *Halictus* and *Lasioglossum* (both referred to as sweat bees), the metallic green *Agapostemon,* the ground-nesting *Dufourea,* and the *Sphecodes,* or cuckoo bees.

SWEAT BEES   **Genus *Halictus***

Common California species:

Halictus tripartitus   `NATIVE`

H. ligatus   `NATIVE`

GENUS SUMMARY: *Halictus* (hah-LICK-tuss) is a widespread and abundant genus that occurs in both the New World and Old World. There are about 200 identified species worldwide and 25 species in North and South America, and seven spe-

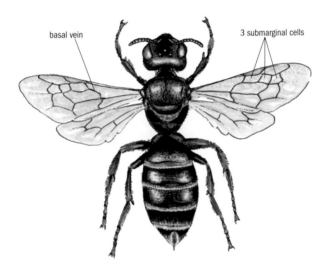

basal vein

3 submarginal cells

Halictus tripartitus, native

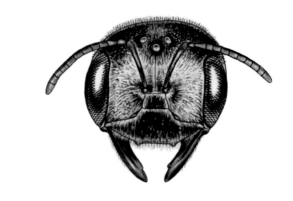

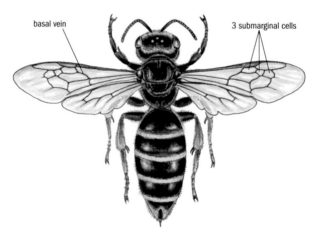

basal vein

3 submarginal cells

Halictus ligatus, native

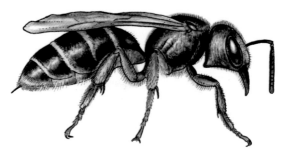

Halictus ligatus, native (profile)

cies are known from California. Each of the species found in California is broadly distributed across the western United States—and in some cases, Mexico and Canada. *Halictus* and the closely related genera *Lasioglossum* and *Agaposte-mon* are all called sweat bees because they are attracted to human sweat and drink it for its salt content.

There is one species in California, *Halictus harmonius,* which is listed on the Xerces Society's Red List of endangered insects. This species is found in the San Bernardino and San Jacinto Mountains.

DESCRIPTION: *Halictus* are small to medium-size bees, 0.2–0.6 in. (4.5–14 mm) long. They are dark brown to black and many species have a dark metallic green sheen. They have bands of hair on the outermost edge of the tergites on their abdomen. Females carry pollen on brushes of hair (scopae) on their hind legs. *Halictus* have three submarginal cells in their fore wing. Under magnification, you can see they have a distinctive wing characteristic, the basal vein strongly arched at the base of the vein; if you look at the submarginal cells of the fore wing, all the cross veins are equally strong, which means you can see both edges of the veins. This is different from what you find in *Lasioglossum*, where the cross veins are weak and look like a single line.

SIMILAR INSECTS: *Lasioglossum* are similar to *Halictus* but can be distinguished by the location of hair bands on the segments of their abdomen and the thickness of the cross veins of the submarginal cells as I mentioned above. *Halictus* has bands of hair on the outer edge of each segment, whereas *Lasioglossum* has bands of hair on the inner edge of each segment.

FOOD RESOURCES: Most *Halictus* are generalist foragers. They use a number of genera of plants, from the Asteraceae to Scrophulariaceae. They are very common on composites (daisy-like disc and ray flowers) in summer and fall.

NESTS: Almost all *Halictus* in North America are semisocial ground nesters. Daughters in social colonies remain in the nest and help care for the young. Nests vary across species from small nests with a single queen and a few workers to nests with multiple queens and hundreds of workers.

FLIGHT SEASON: Individual species can often be seen from spring to early fall. This is due in part to their semisociality. A single nest can be established in spring and continue to reproduce through to fall.

GREEN SWEAT BEES   Genus *Agapostemon*

Common California species: *Agapostemon texanus*   **NATIVE**

GENUS SUMMARY: The genus *Agapostemon* (ag-uh-PAHST-eh-mon) is widespread and abundant throughout North America, with approximately 40 species in total and five species in California. There are approximately 44 identified species worldwide. They are most diverse and abundant in temperate regions and southwestern US deserts. *Agapostemon* are commonly called sweat bees because they are closely related to, and resemble, bees in the *Halictus* and *Lasioglossum* genera. Unlike those bees, however, *Agapostemon* are not attracted to human sweat.

DESCRIPTION: *Agapostemon* are brightly colored metallic green or blue bees. They are medium-size, 0.3–0.6 in. (7–14.5 mm) long. Females of most species in California have a

Agapostemon texanus (female), native

Agapostemon texanus (female), native (profile)

bright metallic green head and thorax and abdomen; some have a black-and-yellow striped abdomen. Females carry pollen on scopal hairs located on their hind legs. Female *Agapostemon* are relatively fast-flying bees. Males generally have a green head and thorax and a black-and-yellow striped abdomen, and they fly much more slowly because they are often searching flowers for females. Under magnification, you would look for a shield-like indentation on the abdomen to confirm that a large bright green bee is an *Agapostemon*, and the face of the propodeum or the back of the thorax is completely encircled by a clearly raised line.

SIMILAR INSECTS: *Osmia* species can also be metallic green but are more robust and carry pollen on the underside of the abdomen instead of on the hind legs. Augochlora species do not have a raised line on their thorax. Some cuckoo wasps also look similar to *Agapostemon*.

FOOD RESOURCES: *Agapostemon* are generalists. Like other members of the family Halictidae, they are short tongued and thus have difficulty extracting nectar from deep flowers. Males can be observed flying slowly around flowers looking for females.

NESTS: *Agapostemon* dig deep vertical burrows in flat or sloping soil, or sometimes in banks. Most species are solitary, but some species nest communally. Up to two dozen females may share a single nest entrance, but each individual builds and provisions its own cluster of brood cells.

Agapostemon texanus (male), native

Where a communal nest gallery shares a single entrance, one bee usually guards the hole, with only her head visible from aboveground.

FLIGHT SEASON: These are summer to fall bees.

SWEAT BEES — Genus *Lasioglossum*

Common California species: *Lasioglossum titusi* `NATIVE`

GENUS SUMMARY: *Lasioglossum* (laz-e-o-GLOSS-um) is a very large genus that occurs worldwide. There are approximately 280 species in North America and approximately 87 species in California. They are often the most common bees in a habitat but are frequently overlooked because of their small size. *Lasioglossum* are closely related to the genera *Halictus* and *Agapostemon*. These genera are commonly called sweat bees because they are known to be attracted to human sweat, which they drink for its salt content.

DESCRIPTION: *Lasioglossum* are slender, dusky black to brown, dull green, or blue bees, with bands of hair on their abdomen. They are tiny to medium-size bees, 0.1–0.4 in. (3–10 mm) long. Females carry dry pollen on scopae (brushes of hair) on their hind legs. Under magnification, you can see that in the fore wing the basal vein is strongly arched at the base of the vein and that the submarginal cells are weak, which means they look like a single line rather than a double line.

SIMILAR INSECTS: *Lasioglossum* are similar to *Halictus* but can be distinguished by the location of hair bands on their abdomen. *Lasioglossum* have bands of hair on the innermost portion of each segment, whereas *Halictus* has bands of hair on the outermost portion of each segment. The wing veins in *Lasioglossum* are also not as thick as those on *Halictus*.

FOOD RESOURCES: *Lasioglossum* are a mix of specialist and generalist foragers. Specialists are known to be attracted to *Oenothera*. At least one species that is a generalist has a strong preference for *Clarkia*. Some are cleptoparasites.

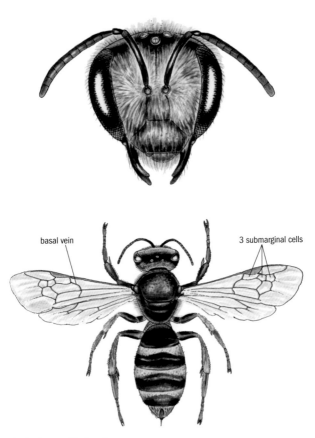

Lasioglossum titusi, native

Lasioglossum titusi, native

NESTS: *Lasioglossum* includes species that exhibit the full range of bee social behaviors, including solitary, communal, and social habits. Most species nest in the ground. In social colonies, daughters remain in the nest and help care for the young. Some of the social species have small nests with a single queen and a few workers, whereas others build long-lived nests with multiple queens and hundreds of workers. Some *Lasioglossum* have glands that produce a mix of chemicals called lactones. Each individual bee has its own unique combination of lactones and lines the nest entrance, burrow, and cells. This helps a worker recognize her own nest when she returns from foraging.

FLIGHT SEASON: These bees are seen from early spring to summer. Some species persist into fall.

SWEAT BEES   Genus *Dufourea*

Common California species: *Dufourea vernalis*   **NATIVE**

GENUS SUMMARY: *Dufourea* (dew-FOUR-ea) is a Holarctic genus that is found all across North America. There are approximately 170 identified species worldwide, approximately 80 species known from North America, and 57 in California.

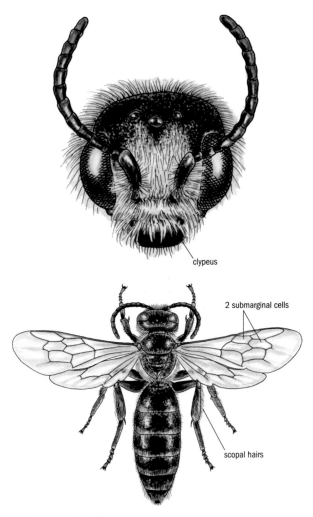

clypeus

2 submarginal cells

scopal hairs

Dufourea vernalis, native

Most species are found only in the western United States, and many are unique to California.

DESCRIPTION: *Dufourea* are small to medium-size, 0.1–0.4 in. (3.5–11 mm) long, narrow-bodied bees. They are usually black, dull green, or metallic blue, and sometimes they have a red abdomen. They may have slight bands of pale hair on their abdomen. The antennae on this genus are positioned very low on the head—below the midpoint of the eyes. Under magnification, you can see the clypeus is usually short and wide, and the labrum is nearly as long as the clypeus. They have two submarginal cells. These bees are diverse morphologically. In particular, they have a lot of variation in their mouthparts. These bees are usually specialists, and the variability in their mouthparts is probably related to the different host plants that each species uses.

SIMILAR INSECTS: *Dufourea* looks like many other Halictidae, particularly members of the genus *Lasioglossum*. The antennae of *Dufourea* are much lower on the head than on *Lasioglossum*.

FOOD RESOURCES: Most of the species in this genus are specialist bees. They specialize on a wide variety of different plants, including *Linanthus, Eschscholzia, Clarkia, Oenothera, Helianthus, Campanula, Calochortus, Monarda, Phacelia, Cryptantha, Mimulus,* Cactaceae, and various others.

NESTS: These are ground-nesting bees. They appear to have fairly shallow nests. These bees are parasitized by another species of bees, and you can sometimes see several parasites hovering above a nest entrance waiting to sneak in and lay an egg in the wall of a cell while it is being provisioned.

FLIGHT SEASON: These bees are found from spring to fall.

CUCKOO BEES   Genus *Sphecodes*

Common California species: *Sphecodes*   NATIVE

GENUS SUMMARY: *Sphecodes* (sfe-KO-deze) is a large cleptoparasitic genus found on all continents and is widespread

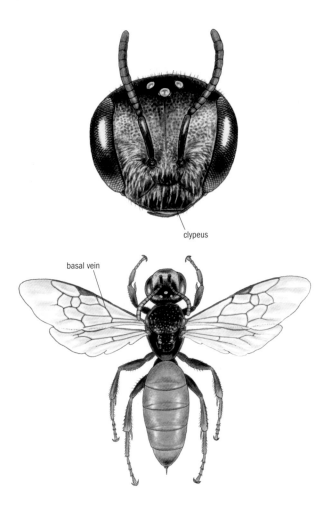

clypeus

basal vein

Sphecodes unidentified species, native

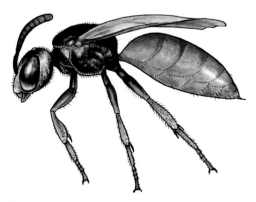

Sphecodes unidentified species, native (profile)

throughout its range. There are about 300 identified species worldwide, 80 species in North America, and it has been estimated there are about 60 species in California, many unidentified and without names. These bees are not very well known and are not well represented in museum collections, so these are only estimates of the numbers of species. *Sphecodes* are cleptoparasites; that is, they invade the nests and lay their eggs in the cells of other species. Their offspring then kill the host larvae and take over the provisions of the host bees. Members of the genus *Sphecodes* primarily lay their eggs in the nests of *Halictus* and *Lasioglossum* species. These cleptoparasites are commonly referred to as cuckoo bees because their behavior mimics that of cuckoo birds.

DESCRIPTION: These are shiny brown to black slender bees with few hairs. They often have red on all or part of the abdomen. They are minute to moderate in size, 0.2–0.6 in. (4.5–15 mm) long. Under magnification, you can see that the basal vein of the fore wing is strongly arched at the base of the vein and all the veins are equally strong. Like all other cleptoparasites, *Sphecodes* lack pollen-collecting hairs because they visit flowers only to drink nectar, not to collect pollen. With a microscope, you can see the relatively large depressions (punctuation) in the exoskeleton. Females have spines on their hind legs that they use to brace themselves in the tunnel of a nest when being attacked by a host bee. They also have lots of ridges and lamellae in their exoskeleton that help deflect the sting and mandibles of an angry host.

Many of the bees in this genus are unnamed or undescribed. What this means is that they have been discovered but have not yet been formally described and named in the scientific literature. Until a formal description is published, a species is considered to be undescribed and unnamed.

SIMILAR INSECTS: *Sphecodes* can resemble wasps. Males of *Sphecodes* look similar to *Lasioglossum* males but have a shorter face and antennae and an entirely black clypeus. They have no hair bands on the terga, unlike *Halictus.*

FOOD RESOURCES: *Sphecodes* do not make their own nests, so they do not forage for pollen for their offspring. They will visit a variety of flowers for nectar.

NESTS: The entire genus *Sphecodes* is cleptoparasitic on many other genera of bees. Cleptoparasites do not construct their own nests. They instead lay their eggs in the nests of other bees, entering a nest while the host female is away gathering pollen. When their larvae hatch, they kill the host bee's larvae or eggs and eat their provisions.

FLIGHT SEASON: These are late spring to early fall bees. There are some early spring records for Southern California.

MINING BEES (Family Andrenidae)

Family Andrenidae is one of the five major bee families with over 2,500 described species. They are widely distributed but absent from Australia, southeastern Asia, and most rain forests. They are medium-size to large bees and are generally dark in color, though some come in blue, yellow, and red. This group does have a unique diagnostic characteristic: the presence of two subantennal sutures (that look like dark lines) under the sockets of the antennae. Next to the antennal sutures are facial foveae. In the Andrenidae, these foveae are covered in hairs and look like patches of velvet when viewed under the microscope. Members of the Andrenidae are usually solitary but may be communal. All of the species in the family Andrenidae nest in the ground, which is why they are often called miner bees. Many species are floral specialists and are common early in spring. Andrenidae usually have a very weak sting, and even females can be safely handled.

There are 12 genera in the family Andrenidae in North America. In California, they are represented by several genera, including *Panurginus* and the very large genera *Andrena* and *Perdita*.

DIGGER BEES or MINER BEES   Genus *Andrena*

Common California species: *Andrena caerulea*   NATIVE

GENUS SUMMARY: *Andrena* (an-DREE-nuh) is a large, well-known genus that occurs in the Americas, Eurasia, and the Old World tropics with over 1,300 species. In California, there are approximately 261 species. This may be the genus with the most species in all of California.

DESCRIPTION: *Andrena* are small to medium-size bees, 0.3–0.7 in. (7–18 mm) in length. They are moderately hairy bees

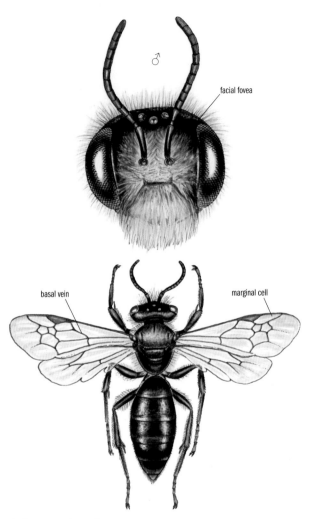

Andrena caerulea, native

Andrena caerulea, native (profile)

that are black or dull metallic blue or green. Most species have pale bands of hair on their abdomen, and the abdomen is long relative to other bee groups. Females have wide facial foveae (depressions) lined with white to brown hairs, which make them seem velvety, and large scopae (brushes of hairs for holding pollen) on the upper part of their legs, seemingly in their "armpits." Under magnification, you can see that males have two subantennal sutures and a straight basal vein in the fore wing. In addition, the marginal cell is pointed and rests on the margin of the wing. Most *Andrena* have three submarginal cells in the fore wing. *Andrena* have a variety of striking colorations, but species are difficult to tell apart from each other.

SIMILAR SPECIES: Some *Halictus* appear similar to *Andrena*. The *Andrena* can easily be differentiated by their facial foveae.

FOOD RESOURCES: The genus contains both generalist and specialist species. Some plant taxa, such as the evening primroses *(Oenothera),* are highly dependent on *Andrena* for pollination. *Andrena chalybaea,* one of our native California species, is considered to be monolectic, meaning it generally collects pollen from only one species, *Camissonia ovata.* This is an unusual degree of specialization.

NESTS: Most *Andrena* are solitary nesters, and they often nest in large aggregations. A few species nest communally, where two or more females share a nest but build and provision their own nest cells. All *Andrena* nest in the ground; they often prefer sandy soil near or under shrubs. Brood cells are lined with a waxy material secreted by the female, but this is absorbed by the surrounding soil rather than becoming like cellophane.

FLIGHT SEASON: These bees fly in early spring. Some species emerge as early as March or April.

PANURGINUS BEES Genus *Panurginus*

Common California species: *Panurginus nigrihirtus*   `NATIVE`

GENUS SUMMARY: *Panurginus* (pan-ur-JINE-us) is a small genus that is fairly uncommon in western North America and less common in the east. There are approximately 47 identified species worldwide and 19 species in North America. In California, there are approximately 14 species. While fairly uncommon as a genus, they are often locally abundant.

DESCRIPTION: *Panurginus* are small, moderately hairy bees, 0.3–0.7 in. (7–18 mm) in length. They are usually jet black, sometimes shiny, with dark-colored hairs. Under magnification, you can see that they have two subantennal sutures, and females have facial foveae that are small and narrow and are not filled with hairs like those of *Andrena.* The clypeus on the face of the males often has yellow marks, as do the legs. The abdomen is long relative to other bee groups. Males are about half the size of females. The pollen-collecting hairs (scopae) are on the hind tibia.

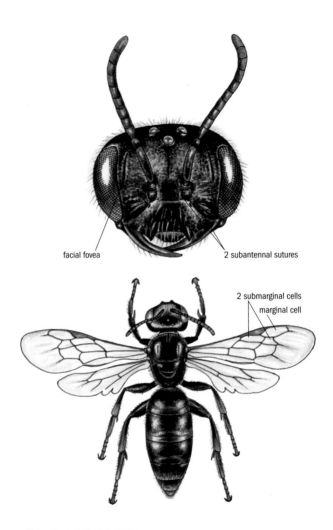

facial fovea

2 subantennal sutures

2 submarginal cells

marginal cell

Panurginus nigrihirtus, native

SIMILAR INSECTS: Some *Halictus* appear similar to *Panurginus,* and there is another genus in the family Andrenidae, *Pseudopanurgus,* which is very similar.

FOOD RESOURCES: The genus contains mostly specialist species, though some are generalists. Even though they are specialists, the different species of the genus use many different flowers. Species can be found on *Ceanothus, Nemophila, Ranunculus,* and other spring native plants.

NESTS: Most *Panurginus* are solitary nesters, though they can sometimes be communal. In other words, in some species, we see both nests occupied by a single female and nests shared by several females. Some species use the same area for nesting year after year, and in some areas they can form large aggregations with over 100 nests per square meter. Nests are often in the top 20 cm of soil. When there is a large nesting aggregation, the males often patrol the nest area looking for females. The males can get so excited that they will often form a "mating ball" by having two to five males mob a newly emerging female (and sometimes a newly emerging male by mistake). The female will escape the ball with one male riding on her back.

FLIGHT SEASON: These bees fly in spring.

MINER BEES   Genus *Perdita*

Common California species: *Perdita rhois*   NATIVE

GENUS SUMMARY: *Perdita* (per-DIH-tuh) is a very large genus that is common in western North America. The center of diversity of this genus is in the southwestern deserts of the United States and northwestern Mexico. There are approximately 629 identified species worldwide, most in North America. In California, there are approximately 253 species.

DESCRIPTION: *Perdita* are very small bees, 0.1–0.4 in. (2–9 mm) in length. They are usually black but sometimes metallic green or blue and have abdominal hair bands. They often

2 subantennal sutures

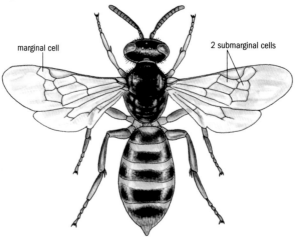

marginal cell

2 submarginal cells

Perdita rhois, native

have yellow or white markings. The profile of their body often seems flat relative to other bees, and their wings have reduced venation. They have two submarginal cells. Under magnification, you can see that there are two subantennal sutures in the face. The marginal cell, which is on the wing edge, is very short; the distance where it borders the margin of the wing is less than half the distance from the tip of the marginal cell to the wing tip.

SIMILAR INSECTS: Some small wasps resemble *Perdita*.

FOOD RESOURCES: The genus contains all specialist species. Even though they are specialists, the members of the genus use many different flowers.

NESTS: Most *Perdita* are solitary ground nesters, though they can sometimes be communal. One unique aspect of a *Perdita* nest is that the females do not line their brood cells. Instead, females cover the pollen ball with a glandular secretion. When a larva hatches, it then chews through this protective covering to reach the pollen.

FLIGHT SEASON: These bees fly from summer to fall.

LEAF-CUTTER BEES or MASON BEES (Family Megachilidae)

This is a large family that includes both mason bees and leaf-cutter bees. While most bees carry pollen on their legs, a distinguishing characteristic of this family is that the non-parasitic female carries pollen on the underside of the abdomen. This can best be observed while the bee is in flight. With a hand lens or microscope, you can recognize a bee as a member of this family because each has a labrum (the upper lip of the mouthparts) that is rectangular and wider than it is long. It also has two (not three) submarginal cells in the fore wing, and the subantennal sutures (lines coming out from the base of the antennae) go outward rather than down. These are long-tongued bees. They primarily nest in preexisting holes.

The family Megachilidae contains some of the most important agricultural pollinators. An introduced species, *Megachile rotundata,* the Alfalfa Leaf-cutter Bee, is an important pollinator of alfalfa. *Osmia lignaria,* the Blue Orchard Bee, is used for pollination of many fruit trees. These bees can be used as pollinators in part because of their ability to nest in artificial nest blocks. A filled nest block can be moved from farm to farm, bringing with it the necessary pollinators. These bees are efficient pollinators because of how they work on flowers. They almost "swim" through flowers, landing right on the reproductive structures of the fruit tree blossoms. And because their pollen-laden scopae are located on the abdomen, they easily transfer pollen from abdomen to anthers as they go, allowing for increased efficiency in pollination.

There are 25 genera of the family Megachilidae in North

America. Some of the common California genera are *Dianthidium, Anthidium, Megachile, Osmia, Ashmeadiella, Hoplitis, and Heriades.*

RESIN BEES or MASON BEES   Genus *Dianthidium*

Common California species: *Dianthidium pudicum*   **NATIVE**

GENUS SUMMARY: The genus *Dianthidium* (die-AN-thid-e-um) is restricted to North America with about 30 species widely distributed throughout California, with 15 species in total. The bees have beautiful pale markings on the abdomen. These bees are known as mason bees because they collect pebbles and glue them together with resin to create their nests. They construct their nests on twigs or on roots below the soil surface or in any available burrows or borings.

DESCRIPTION: These are black or brown bees with pale white, yellow, or ocher patterns on their abdomen. They are small to medium-size bees, 0.2–0.5 in. (5–12 mm) long. They have a cylindrical body shape. They have two submarginal cells in their fore wing. These bees are distinguished under magnification by the shape of their pronotal lobe. It is broad, plate-like, and half-transparent. They also have a spine or tooth on the middle tibia of the hind leg. *Dianthidium* females carry dry pollen in a scopa (brush of hairs) on the underside of the abdomen rather than on their hind legs like most bees. This is characteristic of all females in the family Megachilidae, which also includes *Megachile, Osmia,* and *Hoplitis.*

SIMILAR INSECTS: Bees in the genus are closely related to and resemble *Anthidium.* The members of the genus *Anthidium* are less cylindrical and often larger than the *Dianthidium. Dianthidum* also have an arolium on their tarsus and *Anthidium* do not.

FOOD RESOURCES: *Dianthidium* are generally pollen specialists. Almost all the California species specialize on members of the family Asteraceae. These bees are short tongued.

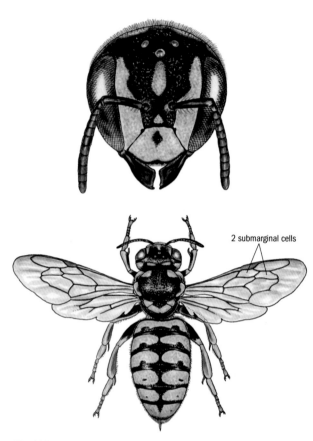

2 submarginal cells

Dianthidium pudicum, native

pronotal lobe

Dianthidium pudicum, native (profile)

NESTS: *Dianthidium* construct their own nests out of pebbles and resin as described above. Most species are solitary. One of the most common California species, *Dianthidium ulkei*, has males that are larger than females and that are very territorial at nest sites.

FLIGHT SEASON: These are primarily late spring to summer bees. In some desert areas, they can be seen as early as the end of March.

CARDER BEES   Genus *Anthidium*

Common California species: *Anthidium mormonum*   **NATIVE**

GENUS SUMMARY: The genus *Anthidium* (AN-thid-e-um) is widely distributed throughout the world with about 170 identified species. There are approximately 29 species in North America and 23 species in California. The bees have beautiful pale markings on the abdomen like the *Dianthidium*. These bees are known as mason bees because they collect pebbles and glue them together with resin to create their nests.

DESCRIPTION: *Anthidium* are robust black bees with a conspicuous yellow or white patterning of stripes that are inter-

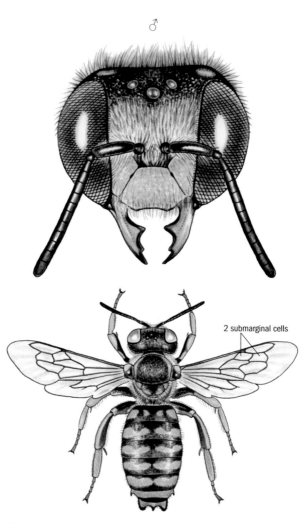

♂

2 submarginal cells

Anthidium mormonum, native

rupted in the middle of their abdomen. They are of moderate size, 0.3–0.8 in. (3–20 mm) long, similar in size to a Western Honey Bee. Their abdomen is broad and somewhat flattened. Like other female bees in the family Megachilidae, *Anthidium* females carry dry pollen in a scopa (brush of hairs) on the underside of the abdomen rather than on their hind legs like most bees. Male *Anthidium* are larger than females. They have two submarginal cells on their fore wing. Using a microscope, you can see that the subantennal sutures are very straight and that each mandible has six or more teeth.

SIMILAR INSECTS: Bees in the genus are closely related to and resemble *Dianthidium*. The members of the genus *Anthidium* are less cylindrical and often larger than the *Dianthidium*. *Anthidium* do not have an arolium and *Dianthidium* do.

FOOD RESOURCES: *Anthidium* are a mix of generalists and specialists. Almost all the California species that specialize forage on flowering legumes. Like other members of the family Megachilidae, they carry pollen on the underside of their abdomen and are short tongued.

NESTS: *Anthidium* find cavity nests in a variety of places. They can nest in the old burrows of other bees or other insects. Sometimes they nest in the stems of plants. Others construct nests from resin and pebbles, placing them on branches of trees. Nesting females of *Anthidium* use the hairs (or "wool") from plants to line their burrows, using their mandibles to "card" the fibers into cell walls. Several cases have been recorded of *Anthidium* bees nesting in keyholes. One species in this genus, *Anthidium manicatum,* is native to Europe and has been introduced to North and South America. Its range is quickly spreading from the eastern United States to the west.

FLIGHT SEASON: These are primarily late spring to summer bees.

LARGE LEAF-CUTTER BEES   Genus *Megachile*

Common California species:

Megachile periherta   **NATIVE**

M. apicalis   **NONNATIVE**

GENUS SUMMARY: *Megachile* (meg-uh-KILE-e) is a large genus that occurs worldwide. There are over 1,500 species in the world, 139 in the United States and Canada, and about 77 in California. The bees are commonly known as leaf-cutter bees because they cut the leaves or flowers of plants and use the pieces to form nest cells. They are the "large leaf-cutter bees," whereas bees of the genus *Osmia* are the "small leaf-cutter bees."

DESCRIPTION: There is wide variety in form and structure among *Megachile* species. Bees in the genus *Megachile* are medium to large bees, 0.4–0.8 in. (10–20 mm) long. Many *Megachile* are dark gray and stout bodied, with a flattened abdomen with pale hair bands but no bands on the integument. *Megachile* females carry dry pollen in a scopa (brush of hairs) on the underside of their abdomen rather than on their hind legs like most bees. This is characteristic of all females in the family Megachilidae, which also includes *Anthidium, Osmia,* and *Hoplitis.* Females of this group often have huge mandibles for cutting leaves. *Megachile* have two submarginal cells on the fore wing. The genus includes the largest bee in the world—a Malaysian species with a 2.5-in. (65 mm) wing span. Under magnification, you can see that they do not have an arolium.

SIMILAR INSECTS: *Megachile* are very distinctive. While other members of the family Megachilidae carry pollen on the underside of their abdomen, they are rarely mistaken for *Megachile* because of the pugnacious body shape of this genus, the pale hair bands on the abdomen, and the way they hold their abdomen when on a flower. They often lift

2 submarginal cells

Megachile periherta, native

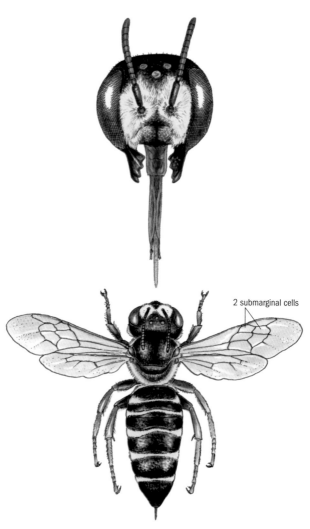

2 submarginal cells

Megachile apicalis, nonnative

Megachile apicalis, nonnative (profile)

their abdomen into the air as they sit on flowers and gather pollen.

FOOD RESOURCES: *Megachile* includes both specialist and generalist foragers. Some species specialize on flowers of plants from the family Asteraceae. *Megachile rotundata*, the Alfalfa Leaf-cutter Bee, is one of the most important agricultural pollinators. It is an introduced species used commercially to pollinate alfalfa.

NESTS: *Megachile* are primarily cavity nesters and nest in a wide variety of habitats and sites, but they all line their brood cells with leaves or petals of plants. Many species are opportunists; their nests have been found in a variety of man-made structures such as garden hoses and crevices in walls. These species prefer to nest in soft rotted wood, abandoned beetle tunnels, and hollow plant stems of large pithy plants (e.g., rose canes, green ash, lilac, Virginia creeper). Other *Megachile* species use resin and mud to build aboveground nests, and a few species nest in the ground. Gardeners do not need to worry; the leaf-cutting and stem-nesting cause little damage to plants. For those species that cut leaves to add to their nest, it only takes 2 or 3 seconds for *Megachile* to cut a piece of leaf. Just before she finishes cutting the leaf, a female *Megachile* starts to beat her wings so that she is already flying by the time the leaf fragment is severed from the plant. In many *Megachile* species, the female flies off with the leaf while gripping its edges and holding it close to her body; she then lands briefly to get a better grip on the piece of leaf, pulling it forward and grasping it between her mandibles and first two pairs of legs for the flight back to her nest.

FLIGHT SEASON: These are primarily summer bees.

SMALL LEAF-CUTTER BEES or MASON BEES
Genus *Osmia*

Common California species:

Osmia nemoris **NATIVE**

O. coloradensis **NATIVE**

GENUS SUMMARY: *Osmia* (OZ-me-yuh) is a large genus found in Eurasia and the New World, and many *Osmia* species are widespread and abundant. There are approximately 350 identified species worldwide, 135 species in North America, and 106 in California. They are most common in most of the western United States but rare in deserts.

DESCRIPTION: *Osmia* are stout, rounded, small to medium-size bees, 0.2–0.8 in. (5–20 mm) long. Most species have a metallic sheen, and many are shiny brilliant metallic green, blue, or even purple. They usually do not have conspicuous bands of hair on their abdomen or markings anywhere on their integument. They have an arolium. They have a broad head and wide abdomen. Under magnification the parapsidal line on the thorax can be hard to see because it is condensed to a simple pit or short line. *Osmia* females carry dry pollen in a scopa (brush of hairs) on the underside of the abdomen rather than on their hind legs like most other bees. This is characteristic of all females in the family Megachilidae, which also includes the genera *Megachile, Anthidium,* and *Hoplitis.*

SIMILAR INSECTS: *Osmia* may be mistaken for some of the other green bees such as *Agapostemon* and *Augochlora.* *Osmia* are generally much more rounded than these species.

FOOD RESOURCES: Most *Osmia* species are generalist foragers and commonly visit flowering shrubs and small trees in the family Rosaceae, especially fruit trees in orchards. Some species prefer monkey flower, lotus, *Phacelia* species, and plants in the families Asteraceae (asters, daisies) and Fabaceae (legumes). They are very efficient pollinators on some fruit trees. For example, one or two hives of Western Honey Bees (approximately 10,000–25,000 foragers) are needed in order to pollinate an acre of apple trees, whereas just 250 female *Osmia* bees can accomplish this same task.

NESTS: *Osmia* are solitary nesters. They nest in a variety of sites and have a diversity of nest architectures. Most *Osmia* do not construct their nests themselves, but rather build cells in preexisting narrow tunnels such as beetle burrows found in woody plants, the pithy or hollow centers of some plant stems, crevices between stones, and abandoned wasp or bee nests. Some are known as leaf-cutter bees because they cut the leaves or flowers of plants and use the pieces to form nest cells. Smaller-bodied *Osmia* species are known as the small leaf-cutter bees, whereas their bulkier counterparts, the *Megachile,* are called large leaf-cutter bees. Other species of *Osmia* are called mason bees because they work

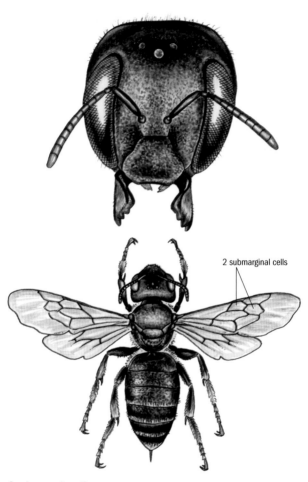

2 submarginal cells

Osmia nemoris, native

Osmia nemoris, native (profile)

mud into partitions to separate their brood cells. One North American species, *Osmia conjuncta,* make their nests in empty snail shells.

Osmia are opportunists; if you provide them with artificial nest sites such as wood blocks with holes, paper drinking straws, or bamboo, they will readily occupy them. *Osmia* will also collect a variety of materials to construct nest cells, such as chewed leaves, mud, and tree resin. Some species use a special pair of horns on their head to pack and smooth mud, whereas others use special spurs on their mandibles. The *Osmia* queen will typically lay eggs that will hatch into male bees in the two cells closest to the entrance, protecting the more important female bees in the back of the nest, away from predators and parasites that may attack the foremost brood cells. The oldest bee in a nest is the first to emerge from its cocoon; it is normally in the deepest cell, farthest from the nest entrance. It proceeds to nip its neighbor, which then emerges and nips its next neighbor, and so

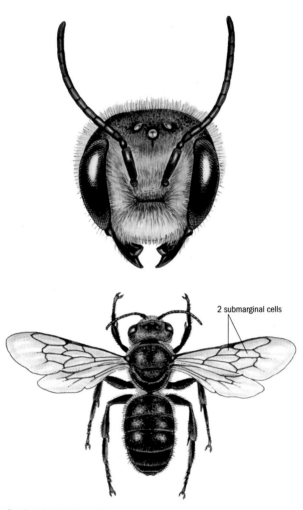

2 submarginal cells

Osmia coloradensis, native

on until the bee in the shallowest cell is awake, thus permitting them all to exit the nest. These behaviors are common in bees with similar nest architectures.

FLIGHT SEASON: These are primarily spring-to-summer bees.

LEAF-CUTTER BEES   Genus *Ashmeadiella*

Common California species:
Ashmeadiella californica

`NATIVE`

GENUS SUMMARY: *Ashmeadiella* (ASH-mead-e-el-uh) are found only in North America from Canada to the Yucatán. There are about 55 species in North America and 41 in California.

DESCRIPTION: These bees are small bees, 0.1–0.4 in. (3.5–9.5 mm) in length. They are stout, nonmetallic bees. They are generally black to brown but sometimes red with bands of pale hair across the abdomen but no markings on the integument. They have two submarginal cells in the fore wing. They have an arolium. These bees are especially abundant in dry areas.

SIMILAR INSECTS: These bees resemble both *Hoplitis* and *Osmia,* other members of the family Megachilidae.

FOOD RESOURCES: This genus has both generalist and specialist bees. The specialists are found on legumes (such as *Dalea* and *Lotus*), on cactus (such as *Opuntia*), and on composites, as well as *Penstemon, Phacelia,* and *Larrea.*

NESTS: *Ashmeadiella* are solitary nesters. They build their nest chambers in holes in wood or stems (including the tip of a cactus stem), in burrows in the ground, or in spaces under rocks. The queens probably mix nectar with leaf pulp to make the divisions between nest chambers. There is a record of an *Ashmeadiella* from Texas nesting in snail shells.

FLIGHT SEASON: Some members of the genus probably produce multiple generations over the course of a year, as they can be observed from early spring to fall. Some species in the

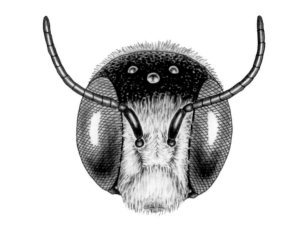

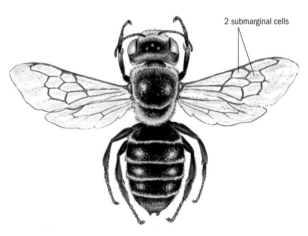

2 submarginal cells

Ashmeadiella californica, native

Ashmeadiella californica, native (profile)

genus have fairly restricted fight seasons, but *A. californica* has been collected from April to November at a single site.

MASON BEES **Genus *Hoplitis***

Common California species: *Hoplitis producta* NATIVE

GENUS SUMMARY: *Hoplitis* (hop-LIE-tuss) are found only from northern Canada south to Mexico as well as in Africa. There are about 95 species in North America and 50 in California.

DESCRIPTION: This is a very diverse genus. These bees are minute to medium bees, 0.1–0.4 in. (3–14 mm) in length. They are heavy-bodied, thick-headed, black, metallic green, or brightly colored green or blue-green bees. There are some species that have a red or red-and-black abdomen. The sides of the abdomen are usually parallel and often have pale hair bands but not integumental bands. They have two submarginal cells in the fore wing. *Hoplitis* have an arolium. Under magnification, you can see that the end flagellomere (subunit) of each antenna is hooked.

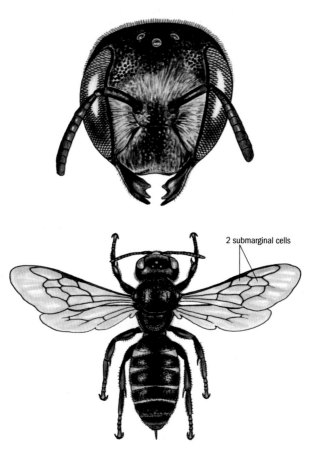

2 submarginal cells

Hoplitis producta, native

SIMILAR INSECTS: These bees resemble *Heriades, Ashmeadiella,* and *Osmia,* other members of the family Megachilidae, but lack the distinctive features of those genera.

FOOD RESOURCES: This genus has a mix of generalist and specialist bees. The specialists are often found on *Lotus, Cryptantha, Larrea, Phacelia,* and *Eriodictyon.*

NESTS: Nests of *Hoplitis* are made in preformed burrows in wood or stems, in holes dug in the pith of plant stems, or in structures made of pebbles and mud. There are even some members of the genus that burrow in the soil. Sometimes they add pebbles and mud to the pith or to the partitions made from chewed leaves. When members of the genus *Hoplitis* lay their eggs, they lay males in the outermost cells and females in the innermost cells.

FLIGHT SEASON: These are summer bees.

MASON BEES   Genus *Heriades*

Common California species: *Heriades occidentalis*   NATIVE

GENUS SUMMARY: *Heriades* (her-EYE-uh-deez) is a large genus of small bees found worldwide except in Australia and most of South America. There are approximately 140 identified species found worldwide. In North America, there are only 11 described species, with four being found in California.

DESCRIPTION: These are small, compact, dark-gray to black bees 0.16–0.28 in. (4–7 mm) in length with long slender bodies. They often have white hair bands on their abdomen but no markings on their integument. Their body surfaces look rough because they have very coarse, dense punctuation (holes in the integument). They lack an arolium.

SIMILAR INSECTS: These bees resemble *Ashmeadiella, Osmia,* and *Hoplitis,* and other members of the family Megachilidae. *Heriades* look much rougher because of their punctuation than these other genera.

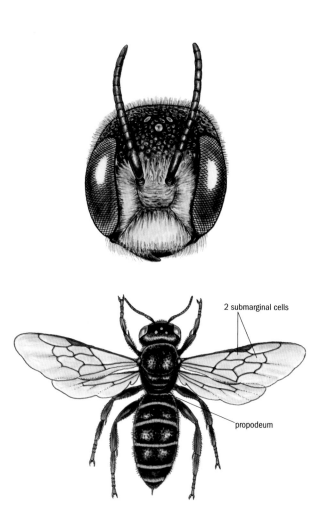

Heriades occidentalis, native

Heriades occidentalis, native (profile)

FOOD RESOURCES: There are both generalist and specialist species in this genus. *Heriades cressoni*, which is found in the high montane and alpine regions of California and along the coast, is a specialist on summer composites. *Heriades occidentalis*, which is more broadly distributed, is a generalist.

NESTS: These are cavity-nesting bees. They generally nest in wood, not stems. One of the California species, *Heriades occidentalis*, has been documented to nest in pine cones.

FLIGHT SEASON: *Heriades* are summer bees.

CUCKOO, CARPENTER, DIGGER, BUMBLE, and HONEY BEES (Family Apidae)

This is the largest of the bee families. It includes many favorite and most well-known bees: honey bees, bumble bees, squash bees, and even some cuckoo bees. This group has a wide variety of species. Most are solitary with simple nests. However, this family includes the most advanced eusocial bees and many of our important agricultural pollinators. Almost 20 percent of the family Apidae are thought to be cleptoparasites, laying their eggs in the nest cells of other bees. There are approximately 84 genera of the family Apidae in North America.

HONEY BEES
Genus *Apis*

Common California species: *Apis mellifera*   NONNATIVE

GENUS SUMMARY: There are approximately seven species of *Apis* (A-pus) in the world. In North America, the genus *Apis* is represented by only one species, the Western Honey Bee. The Western Honey Bee is the most important agricultural pollinator.

DESCRIPTION: *Apis* are moderately hairy, elongated bees with hairy eyes. They are medium-size bees, 0.4–0.6 in. (10–15 mm) long. They vary in color from very dark brown or black to pale gold with darker stripes on their abdomen. Females have corbiculae on their rear legs for carrying moistened pollen. Males are almost never seen.

SIMILAR INSECTS: These bees look similar to *Peponapis* and *Xenoglossa* but are not usually found on squash flowers. They also tend to forage later in the day than other bees do.

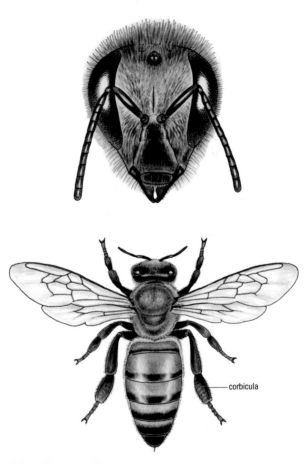

Apis mellifera, nonnative

corbicula

Apis mellifera, nonnative (profile)

FOOD RESOURCES: These bees are generalists. They are used all over the world as pollinators of a variety of crops.

NESTS: *Apis* are cavity nesters. They can be found nesting in spaces ranging from rock crevices to tree holes to walls in buildings to commercial hives. A single colony consists of a queen bee (a fertile female), male bees called drones, and thousands of female worker bees. Within a nest, workers build cells out of wax to create multicelled honeycombs. The queen lays a single egg in a cell. The sex of the resulting bee depends on whether the queen lays a fertilized egg (female) or unfertilized egg (drone). The larvae then are fed by the workers. Larvae that are fed a special diet of royal jelly become queens. Other larvae that are fed royal jelly and then simply honey and pollen become female worker bees. The mature larva spins a cocoon and pupates in the cell.

Within the nest, there are different bees doing different jobs. The youngest bees are responsible for feeding larvae and cleaning the hive. As they age, they move to taking nectar and pollen from foraging bees and guarding the hive and then to being forager bees. This highly organized social structure and elaborate division of labor within the hive is one of the reasons these bees are considered eusocial.

A single hive can house 20,000–30,000 bees in winter and 60,000–80,000 bees in summer. Honey bees are very useful for agriculture because the simple act of moving one hive can increase the number of pollinators in an area by upwards of 60,000.

FLIGHT SEASON: Honey bees are found year-round. In areas with cold winters, they will stay in their hives during frigid periods. In warmer climates, they can be seen year-round.

SQUASH BEES | Genus *Peponapis*

Common California species: *Peponapis pruinosa* NATIVE

GENUS SUMMARY: There are about 13 species of *Peponapis* (pep-on-AY-pus) in the world. There are two species in California. The more common species, *Peponapis pruinosa*, tends to occur in the Central Valley and through all Southern California. It is strongly associated with agricultural areas.

DESCRIPTION: *Peponapis* are fairly robust, hairy, medium-size bees, 0.4–0.6 in. (10–15 mm) long. Squash bees are typically brown with bands of brown hair on their abdomen. They have a large clypeus, which makes them look like they have a big nose. Under magnification, you can see that the base of the mandible is dark and that there is a distinct notch or tooth at the tip of the end of the mandible. This notch is relatively shallow, and in older individuals the mandible can be worn away so much that the notch is gone. The interiors of the fore wing cells have microscopic hairs scattered across the surface. Males have a small yellow patch on their face and very short antennae. These are early morning bees.

SIMILAR INSECTS: There is a very similar genus that also pollinates squash blossoms called *Xenoglossa*. They have the same size and color patterns of Western Honey Bees but can be distinguished at flowers because they arrive about an hour before any Western Honey Bees and just dive right into the corollas. Western Honey Bees are much more likely to hover before entering.

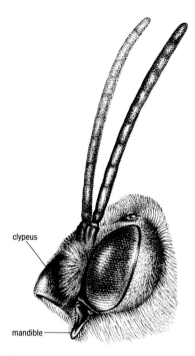

clypeus

mandible

Peponapis pruinosa, native

FOOD RESOURCES: These bees are specialists. They pollinate squash, pumpkins, melons, cucumbers, and other members of the family Cucurbitaceae. *Peponapis* bees are active in the early morning hours because Cucurbitaceae pollen is available only at that time. *Peponapis* are considered to be better pollinators of Cucurbitaceae than European Honey Bees because they transfer more pollen during a single visit.

Since *Peponapis* depend on the pollen of the plant genus *Cucurbita*, they have probably moved into the eastern United States only after squash and pumpkins were brought there for cultivation because there are no native *Cucurbita* in that region.

NESTS: These bees nest in the ground, often right under the plants that they pollinate. During the day, if you squeeze a squash flower, you can often find sleeping male bees inside. Be prepared for the buzz though; it always makes us jump!

FLIGHT SEASON: These are midsummer bees. They are flying when squash plants are in bloom.

HABROPODA BEES   Genus *Habropoda*

Common California species: *Habropoda depressa*   `NATIVE`

GENUS SUMMARY: There are approximately 55 identified species worldwide and 10 species of *Habropoda* (hab-ro-PODA) in California. These bees look very similar to their close relatives in the genus *Anthophora*. They are important pollinators, particularly of blueberries *(Vaccinium)*. Most of the North American members of the genus are found in California.

DESCRIPTION: *Habropoda* are robust, hairy, fast-flying bees. They range in size and are 0.4–0.7 in. (10–18 mm) long. Most are gray or black. Some have hair bands, and some have bands in the outer covering of the body (integument).

SIMILAR INSECTS: These bees look very similar to members of the genus *Anthophora*. Under magnification, you can see that they have an elongate marginal cell in the fore wing that extends far beyond the submarginal cells, and the third submarginal cell is narrowed toward the front.

FOOD RESOURCES: These are a mix of generalists and specialists. *Habropoda* are known to forage on *Arctostaphylos* (including one specialist manzanita species), *Ceanothus, Lupinus, Phacelia,* and *Salvia.*

NESTS: These are ground-nesting bees. They nest either singly or in an aggregation. These bees metamorphose in fall and overwinter as resting adults. Nests of *Habropoda pallida* have been measured as deep as 6 ft (2 m). Their nests are also parasitized by beetles of the family Meloidae.

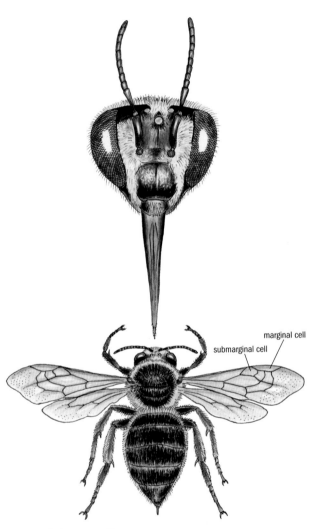

marginal cell

submarginal cell

Habropoda depressa, native

Habropoda depressa, native (profile)

FLIGHT SEASON: These are spring bees. They are active as early as February in most of California and have been collected as late as December in the deserts.

CUCKOO BEES
Genus *Nomada*

Common California species: *Nomada edwardsii* NATIVE

GENUS SUMMARY: There are approximately 700 species worldwide, 287 in North America, and 75 species of *Nomada* (no-MOD-uh) in California. While we can easily recognize the genus, the species are extraordinarily difficult to tell apart. It is even difficult to match up males and females within the same species.

DESCRIPTION: *Nomada* are slender, sparsely haired, and wasplike. They are generally small to medium sized, 0.1–0.6 in. (3–15 mm) long. Most species are black or red, and the majority of species have extensive cream, yellow, or red markings. If the integument is black, then yellow markings are always present. Their antennae appear thick in comparison to other bees. Because they do not visit flowers to collect pollen, *Nomada* lack many of the traits associated with pollen-collecting bees, such as scopae. Under magnification, you can see that they have a distinctive wing characteristic:

vannal lobe

Nomada edwardsii, native

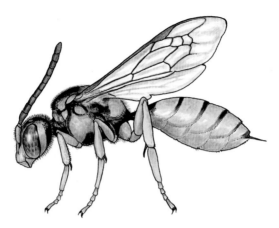

Nomada edwardsii, native (profile)

the jugal lobe of the hind wing is small, approximately one-sixth as long as the vannal lobe or less.

SIMILAR INSECTS: *Nomada* may easily be mistaken for small wasps of the family Sphecidae, but there are a couple of distinguishing characteristics. Sphecidae have silver or gold hairs on the lower part of their face that appear to glitter in the light. *Nomada,* in contrast, have branched (feathery) hairs that do not reflect light.

FOOD RESOURCES: While *Nomada* do not visit flowers to collect pollen, because their offspring usurp the pollen stores of their host species, they do use nectar from a wide variety of flowers.

NESTS: *Nomada* are all cleptoparasites. They most commonly lay their eggs in the nests of members of the genus *Andrena,* but also *Agapostemon, Eucera, Halictus,* and *Lasioglossum. Nomada* typically lay two to four eggs alongside the one egg found in each host brood cell. The first *Nomada* larva to emerge kills its siblings and the host egg, then consumes the cell provision. Young *Nomada* larvae have large sickle-shaped mouthparts that they use to kill their hosts' larvae. Male *Nomada* bees may patrol *Andrena* nest sites, mimicking the odor of *Andrena* females. This behavior attracts *Nomada* females, both to mate and to a potential nest site.

FLIGHT SEASON: These bees track the timing of their host bees and are primarily found in spring.

MINING BEES — Genus *Anthophora*

Common California species: *Anthophora urbana* **NATIVE**

GENUS SUMMARY: *Anthophora* (an-THAH-for-uh) is a large genus of fast-flying bees found throughout the temperate and tropical regions of the world with the exception of Madagascar and Indonesia. There are about 460 identified species worldwide. It is abundant in the United States, and there are about 70 species, primarily in the western states. There are 41 species in California.

DESCRIPTION: *Anthophora* are fast-flying, chunky, and hairy bees. They are small to moderately large bees, 0.2–1.1 in. (5–28 mm) long. Many species have pale hair bands on the abdomen. Their legs appear to be very hairy, but the "hair" is actually their large pollen-carrying scopae. Under magnification, you can see that the interiors of the fore wing cells do not have any microscopic hairs scattered across the surface. In the fore wing, the upper and lower margins of the third submarginal cell are about equal in length. You often hear these bees before you see them.

SIMILAR INSECTS: *Anthophora bomboides* superficially resembles a bumble bee. It also resemble *Diadasia,* but rather than having a rounded head, its head is flat across the top.

FOOD RESOURCES: *Anthophora* are floral generalists with very long tongues (0.4–0.8 in.) that enable them to pollinate deep flowers.

NESTS: All the California *Anthophora,* except one, nest in the ground. *Anthophora furcata* nests in wood. The solitary ground nesters dig tunnels down into soil banks or flat ground. Some species facilitate excavation by wetting clay with regurgitated nectar and water. Brood cells are lined with an oily substance that is almost completely waterproof.

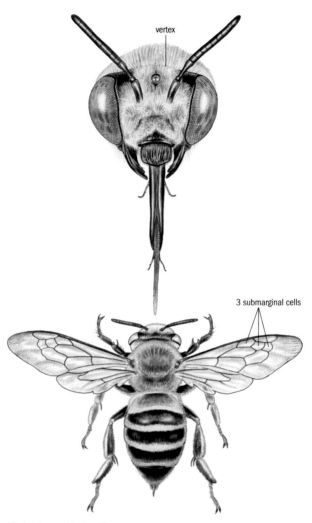

vertex

3 submarginal cells

Anthophora urbana, native

Anthophora urbana, native (profile)

Anthophora do not make a cocoon as pupae, but instead they overwinter as late larval instars or adults. *Anthophora* males often form "sleeping aggregations" in which several males gather on a single plant in the evening and fall asleep holding onto a plant with their jaws. Along the coast of California, there is a species, *Anthophora bomboides,* that mimics bumble bees. It nests in aggregations of hundreds of bees in dune areas and often surprises beachgoers.

FLIGHT SEASON: These are spring and summer bees though some species persist into fall.

BUMBLE BEES   Genus *Bombus*

Common California species:

Bombus vosnesenskii   `NATIVE`

B. californicus   `NATIVE`

B. melanopygus   `NATIVE`

B. flavifrons   `NATIVE`

GENUS SUMMARY: All bees commonly known as bumble bees are in the genus *Bombus* (BOM-bus). There are approximately 250 species of *Bombus* in the world, 40 species

Bombus vosnesenskii, native

Bombus vosnesenskii, native (profile)

in North America, and approximately 26 in California. *Bombus* are primarily found in temperate climates and are particularly important pollinators in alpine and arctic environments. There is a subgenus of *Bombus* called *Psithyrus.* Its bees are social parasites on other *Bombus* species. A *Psithyrus* queen enters the nest of a nonparasitic *Bombus* and kills the queen. The workers of the raided colony then provide for the *Psithyrus* female and her offspring.

DESCRIPTION: *Bombus* are very hairy and very large with yellow, black, orange, or red bands. Both the thorax and abdomen are covered in hair. Under magnification, you can see that these bees do not have a jugal lobe on their hind wing. They have several physiological adaptations that allow them to fly in cold or cloudy weather. They range in length from 0.4 to 0.9 in. (10–23 mm). Species in this genus have very similar body shapes but can be distinguished by their color patterns. Members of the *Bombus* subgenus *Psithyrus* lack pollen-gathering structures because they do not visit flowers but instead steal pollen from other *Bombus* species. *Bombus* females carry pollen wetted with nectar on their corbicula, a smooth, bowl-shaped structure ringed with

Bombus californicus, native

long hairs on the upper part of their hind leg. Bumble bees make a low buzzing sound when flying, usually in a characteristic awkward or "bumbling" pattern.

The species are often told apart by the color patterns of their hair, but even this can vary within a species, so other traits are also used, such as face length and the placement of ocelli. Male bumble bees are even harder to identify than females, because color patterns are even less reliable for males. There are two excellent resources for identifying bumble bees: *Bumble Bees of the Western United States,* by Koch, Strange, and Williams, and *Bumble Bees and Cuckoo Bumble Bees of California*, by Thorp, Horning, and Dunning. In the late 1990s, bee taxonomists began to notice a decline in the abundance and distribution of several bumble bee species, including many bumble bees that were formerly among the most common species in North America. A number of threats face bumble bees that may be leading to the decline of these species, including the spread of pests and diseases through commercial bumble bee rearing and transportation, habitat destruction or alteration, pesticide use, invasive species, and climate change. Bumble bee expert Dr. Robbin Thorp (Professor Emeritus, UC Davis) has hypothesized that wild populations of the closely related species *Bombus occidentalis, B. affinis, B. terricola,* and *B. franklini* were infected by an introduced disease carried by commercially reared colonies of *B. occidentalis* and *B. impatiens.* Two California *Bombus* species are on the Xerces Society's Red List of Bees. Franklin's Bumble Bee *(Bombus franklini)* is listed as "critically imperiled" and "possibly extinct." The Western Bumble Bee *(Bombus occidentalis)* is listed as "imperiled" and is in sharp decline.

The Western Bumble Bee was once very common in the western United States and western Canada. It was arguably the most common bumble bee in San Francisco until the 1990s. Our recent surveys of San Francisco were completely unable to locate this species. *B. occidentalis* can still

Bombus melanopygus, native

be found in the northern and eastern parts of their historic range, but the once common populations from southern British Columbia to central California have disappeared. This bumble bee is an excellent pollinator of greenhouse tomatoes and cranberries and has been commercially reared to pollinate these crops. In the past, it has also been an important pollinator of alfalfa, avocados, apples, cherries, blackberries, and blueberries.

SIMILAR INSECTS: A number of species imitate bumble bees, including other bees, hawk moths, flies, and even beetles. Bumble bees are similar to *Xylocopa,* carpenter bees, but they are generally furrier.

FOOD RESOURCES: *Bombus* are among the first bees to emerge in spring and the last to die off. The large queens are often found foraging on early-blooming willows and *Ceanothus*. They are generalist foragers and visit a succession of flowers throughout the flowering season. Many *Bombus* species have long tongues that enable them to access nectar from deep flowers such as monkshood, foxglove, and lousewort. Some *Bombus* species are managed for the pollination of crops. *Bombus* are more efficient pollinators than Honey Bees *(Apis mellifera)* for many crops, including tomatoes, cranberries, raspberries, blueberries, and field beans. For some species like tomatoes, the pollen is stored in the anthers until the anthers are vibrated at particular frequencies (middle C will work). When the anther is vibrated, it releases its pollen. Honey Bees are unable to vibrate, whereas bumble bees do it with facility. Bumble bees can also pollinate for more months of the year because they are able to work under cooler temperatures than *Apis mellifera.*

NESTS: *Bombus* nest socially in annual colonies that they build in abandoned rodent nests, grass tussocks, or other premade cavities. Nests of some species grow to hold over 1,000 individual bees; however, it is more common for nests to have fewer than 50 bees. Queens overwinter alone and establish a new nest in spring. *Bombus* brood cells are

Bombus flavifrons, native

GUIDE TO SOME OF THE MORE COMMON
BUMBLE BEES OF CALIFORNIA

This is a quick guide to some of the most common color patterns of female bumble bees in California. There is a lot of variation in color patterns in these species, so this guide often will not work outside of California. For a more comprehensive key, see the 2012 book *Bumble Bees of the Western United States* by Koch, Strange, and Williams (USDA Forest Service Research Notes. Publication No. FS-972).

Note: The colors referred to in the key are the colors of the hairs, not the colors of the integument.

Bumble bees with a yellow band behind the head and at the tip of the abdomen
- With a yellow face: *Bombus vosnesenskii* (common), *B. caliginosus* (uncommon), *B. vandykei* (uncommon)
- With a black face: *Bombus californicus*

Bumble bees with a yellow band behind the head and in the middle of the abdomen, with a black or gray face: *Bombus crotchii* (uncommon)

Bumble bees with a mix of yellow and black hairs on the abdomen but primarily yellow toward the thorax and black or red toward the tip of the abdomen: *Bombus flavifrons*

Bumble bees with a yellow face, a yellow band that crosses from the thorax to the abdomen, and a wide band of yellow toward the end of the abdomen
- With some black intermixed on the face: *Bombus melanopygus* (common)
- With a black band on the thorax dividing patches of yellow: *Bombus sylvicola* (uncommon, found in eastern California)
- With a black band on the thorax in the area near the head and a mix of yellow and black or white and black hairs: *Bombus bifarius, Bombus huntii*

unique among all bees; they are closed and grow with the larvae. Sometimes there is more than one larva within each cell, and the cell buds as the larvae mature. The queen forages for a short time to collect provisions for her first brood. Once this first brood of workers is ready to take over foraging, the queen remains in the nest to lay eggs.

FLIGHT SEASON: Queens fly in spring. Workers fly in spring, summer, and fall, and new queens and males fly in the late summer and fall.

LONG-HORNED BEES or SUNFLOWER BEES

Genus *Diadasia*

Common California species: *Diadasia bituberculata*   NATIVE

GENUS SUMMARY: *Diadasia* (die-uh-DAY-zee-uh) is a very diverse genus restricted to the New World. About 40 species occur in North America, mostly in the western states. There are approximately 19 known in California. They (and other genera in the tribe Eucerini) are commonly called long-horned bees because males have exceptionally long antennae. *Diadasia* usually fly in the mornings and early afternoons.

DESCRIPTION: *Diadasia* is a diverse species. Most species are robust and hairy, with distinct bands of pale hair on their abdomen, while others have a uniform covering of pale hair. They are of small to medium size, 0.2–0.9 in. (5–23 mm) long. *Diadasia* females carry pollen in huge scopae on their hind legs. Males have very long antennae and projections on the last tergum, which you can see under magnification. If you catch a *Diadasia* and look at it, you can see that the head is rounded on the top and narrower than the tip of the thorax.

SIMILAR INSECTS: *Diadasia* superficially resemble other genera in the tribe Eucerini, such as *Peponapis*, *Melissodes*, and *Svastra*.

Diadasia bituberculata, native

FOOD RESOURCES: Many *Diadasia* species are specialist foragers on particular plant taxa, primarily on the Malvaceae. They are also found on the genus *Helianthus* and families Cactaceae and Onagraceae. They are referred to as sunflower bees because some species are important pollinators of hybrid sunflower crops. A study from California's Central Valley found that the presence of *Diadasia* bees greatly enhanced the effectiveness of Honey Bees *(Apis mellifera)* as pollinators. Honey Bees and native bees avoid each other, and as a result Honey Bees move more often between the rows of male (pollen-producing) and female (hybrid seed–producing) flowers. When native bees are not present, Western Honey Bees tend to stay in a row, transferring pollen from male flower to male flower and not pollinating the female flowers.

NESTS: *Diadasia* are solitary nesters, and some species often nest in large aggregations. *Diadasia* make relatively shallow nests in the ground, often with turreted nest entrances. *Diadasia* may soften the earth with nectar in preparation for nest excavation.

Diadasia species are important pollinators of hybrid sunflower crops. Pollination rates by *Diadasia* and other native bees decline with increasing distance of crops from natural nesting habitat. Thus, intensification of agriculture and elimination of nearby habitat can be detrimental to populations of *Diadasia* species and other native bees

FLIGHT SEASON: These bees fly primarily in spring.

LONG-HORNED BEES   Genus *Melissodes*

Common California species: *Melissodes lupina*   NATIVE

GENUS SUMMARY: *Melissodes* (mel-ih-SO-deez) is the most abundant genus, in both numbers and number of species, of long-horned bees in California. This is a large genus of common bees restricted to the New World. There are about 130

Melissodes lupina, native

Melissodes lupina, native (profile)

identified species worldwide, 119 species in North America, and about 53 in California. *Melissodes* tend to fly in mornings and early afternoons. Both the name *Melissa* and the name *Melissodes* come from the Greek word for *honey bee* (μέλισσα, or *melissa*) and the word for *honey* (μέλι, or *meli*).

DESCRIPTION: *Melissodes* are small to medium-size bees that are 0.3–0.7 in. (8–18 mm) long. They are robust, hairy bees with very hairy legs. The clypeus on the male is often yellow. Unlike on *Peponapis* and *Eucera*, the clypeus bulges a little bit when viewed in profile. Under magnification, you can see that the interior cells of the wings have tiny hairs. The female carries dry pollen on scopae (brushes of hair) that are feathery on her hind legs. Males also have hairs on their hind legs that can accumulate pollen; however, these hairs are not really designed for carrying pollen. Like other genera in the tribe Eucerini, such as *Eucera*, males of *Melissodes* often have very long antennae, thus earning them the common name "long-horned bees." Unlike other long antennae, *Melissodes* antennae are dark on top and lighter on the underside.

SIMILAR INSECTS: *Melissodes* are very similar to *Svastra*, though those bees are usually larger. *Melissodes* are also confused with other bees in the tribe Eucerini, which includes *Eucera* and *Tetraloniella*.

FOOD RESOURCES: Many of these bees are specialists on the plant family Asteraceae, which includes asters, daisies, and sunflowers. These are the bees that you find on fall composites. Many Asteraceae, such as sunflowers, are highly dependent on *Melissodes* for cross-pollination.

NESTS: Most *Melissodes* are solitary ground nesters. Some species nest communally, with several individuals using one burrow. They line their brood cells with a waxlike material they secrete. *Melissodes* are parasitized by the cuckoo bee subgenus *Triepeolus*. The *Triepeolus* female enters the *Melissodes* nest and lays her own egg in a *Melissodes* brood cell. The *Triepeolus* larva then hatches and eats the provisions that the *Triepeolus* female had gathered for her own young.

FLIGHT SEASON: These bees fly from midsummer to fall.

LARGE CARPENTER BEES   Genus *Xylocopa*

Common California species:

Xylocopa tabaniformis   `NATIVE`

X. californica   `NATIVE`

GENUS SUMMARY: The genus *Xylocopa* (zile-low-COPE-uh) is common and widespread across the United States and in California. There are approximately 400 species worldwide, 32 species in North America, and 3 species in California. These are the largest bees in the state. The males can be territorial and may buzz around humans but rarely sting.

DESCRIPTION: *Xylocopa* are robust with a round head and often with a hairy thorax. These bees are medium-size to large, 0.5–1.2 in. (13–30 mm) long. Most species are black, dark metallic blue, or green with dark wings. Males are sometimes a dark gold. Females carry dry pollen on scopae on their legs but do not have a corbicula. They have strong

marginal cell

Xylocopa tabaniformis, native

Xylocopa tabaniformis, native (profile)

and sharp mandibles, which help with excavating nests. Their wings lack a stigma. Some male *Xylocopa* use scents to attract females to their territories.

SIMILAR INSECTS: *Xylocopa* are often confused with bumble bees *(Bombus)* because they are both very large. However, *Xylocopa* are much shinier and less hairy than bumble bees, and their abdomens, in particular, are shiny in comparison. If you see an all-black large bee, it is likely to be a *Xylocopa*, as there are no black bumble bees in California. Under magnification, you can see that the marginal cell of the fore wing is slender. It is usually about seven times as long as it is broad and is just a bit over half as wide as the marginal cells. They also do not have a stigma on the fore wing. When you look at their hind leg, there are hairs for carrying pollen but no corbicula, unlike a Western Honey Bee.

FOOD RESOURCES: *Xylocopa* visit a wide variety of flowers for nectar and pollen. They tend to visit larger, open flowers because of their large body size. These bees are often thieves. They rob flowers of nectar by chewing holes in the side of the

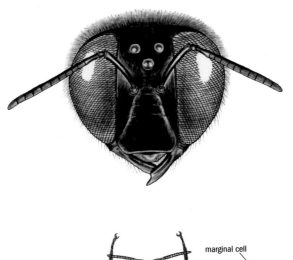

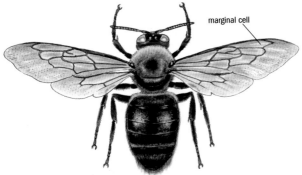

marginal cell

Xylocopa californica, native

base of a flower and sucking nectar out with their mouth-parts through the hole. This means that they do not contact the anthers and stigma at all, so no pollen is transferred. You can often see the slits left by the carpenter bees on the side of tubular flowers. Even though they can rob nectar, *Xylocopa* also pollinate many crops.

NESTS: Most *Xylocopa* make solitary nests in live or dead wood or inside the hollow vertical stems of yucca and agave plants. They burrow into pithy stems of wood, usually soft conifers. Some *Xylocopa* species are semisocial. In these species, bee generations overlap and mothers and daughters share a nest. For some species, this cohabitation is only temporary. For others it persists through adulthood. *Xylocopa* are well known for chewing nest cavities in decks and the walls of buildings. Carpenter bee males patrol their nest areas vigorously. They will even attempt to chase away a human. Fortunately, male bees do not have stingers, so while you may be surprised; you have nothing to worry about!

FLIGHT SEASON: *Xylocopa* fly from early spring to fall in low-elevation areas, June to August in higher elevations and the northern Great Basin. They overwinter as adults.

SMALL CARPENTER BEES   Genus *Ceratina*

Common California species: *Ceratina acantha*   NATIVE

GENUS SUMMARY: The genus *Ceratina* (ser-uh-TEA-nuh) is found on all continents except Antarctica. There are 351 identified species worldwide, 21 species of *Ceratina* that are common in the United States and Canada, and 13 species in California. *Ceratina* is rare in desert habitats. This group is unusual because it has several species that are parthenogenetic: females can produce eggs without mating. One of the best known parthenogenetic species is *Ceratina dallatorreana,* a species native to Europe that has been introduced to

Ceratina acantha, native

Ceratina acantha, native (profile)

California. While male *C. dallatorreana* have been found in Europe, only females have been found in California.

DESCRIPTION: *Ceratina* are long, shiny, glabrous, and dark black, blue, or green. They are closely related to the genus *Xylocopa* (the large carpenter bees), but they are much smaller and flatter. *Ceratina* are small bees, 0.1–0.6 in. (2.5–15 mm) long. Most species have yellow or white markings on the face, and females often have a yellow bar in the middle of their face below their antennae. Some have yellow on their pronotal lobes. Females carry dry pollen on scopae on their hind legs. The tip of a female's abdomen is shaped like a shield. While females have scopae on the backs of their legs, they usually have very small pollen loads.

SIMILAR INSECTS: This genus may be confused with shiny dark bees in the family Halictidae, but *Ceratina* is distinguished by the lack of pollen hairs on its femur (the part of the leg closest to the bee's body).

FOOD RESOURCES: Most *Ceratina* species are generalist foragers and are found on a wide variety of plant species.

NESTS: *Ceratina* includes solitary species and social species. While commonly called carpenter bees, they usually excavate their nests with their mandibles in the pithy centers of dead stems rather than in wood. Females will overwinter as adults in excavated stems and then modify this winter nest into a brood nest by excavating it further. Some of the shrubs they commonly use for nests are elderberry, box elder, sumac, *Rubus* spp., and *Helianthus* spp.

FLIGHT SEASON: These are spring to summer bees. Some species may have two broods in a year.

LONG-HORNED BEES · Genus *Eucera*

Common California species:
Eucera frater albopilosa

`NATIVE`

GENUS SUMMARY: *Eucera* (YOU-sur-uh) is a large Holarctic distinctive genus of bees. They are moderately common. There are about 218 identified species worldwide, 60 species in North America, and 25 species known from California.

DESCRIPTION: These are robust, densely hairy dark bees that resemble the genus *Melissodes*. They have hair bands on their abdomen that contribute to their fuzzy appearance. These are large bees, 0.4–0.7 in. (9–19 mm) in length. The clypeus is distinctive because it is prominent. Under magnification, you can see that their scopal hairs are simple and do not branch. Many males have extremely long dark-colored antennae. The interiors of fore wing cells have microscopic hairs scattered across the surface.

SIMILAR INSECTS: The males of this species can be mistaken for males of the genus *Peponapis,* and the females can resemble the closely related genera *Tetralonia* and *Melissodes.* For all of these comparisons, you need a microscope and pinned specimens to tell them apart. Unlike *Melissodes*, these are most common in spring.

FOOD RESOURCES: *Eucera* are a mix of generalist and specialist bees. Most of the California species are generalists using a range of spring-flowering plant species. The specialists are found on *Arctostaphylos, Amsinckia, Oenothera,* and *Clarkia.*

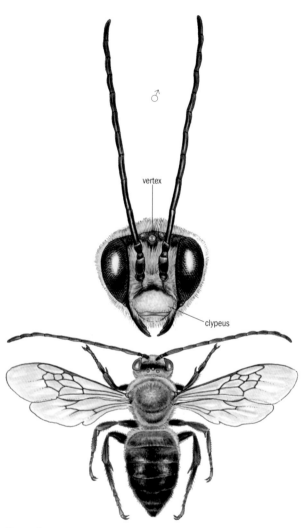

Eucera frater albopilosa, native

Eucera frater albopilosa, native (profile)

NESTS: *Eucera* are solitary ground nesters. They are often found nesting in sandy or clay banks. The nest entrance can often be identified by the presence of a pile of clay or a tumulus.

FLIGHT SEASON: These are exclusively spring bees.

LONG-HORNED BEES   **Genus *Tetraloniella***

Common California species: *Tetraloniella pomonae*   `NATIVE`

GENUS SUMMARY: There are approximately 117 identified species worldwide, 19 species of *Tetraloniella* (tet-ruh-LOAN-e-el-uh) in North America, and 5 species in California. Most of these are very rare in collections; however, *Tetraloniella pomonae* can be locally abundant in California.

DESCRIPTION: These are moderate-size to large fuzzy bees, 0.3–0.6 in. (7–14 mm) long. They strongly resemble *Melissodes* but fly in spring rather than in summer or fall. Their clypeus does not bulge. Under magnification, you can see that their mandible is simple, which means that it does not have any teeth or notches on the upper or lower edges. The

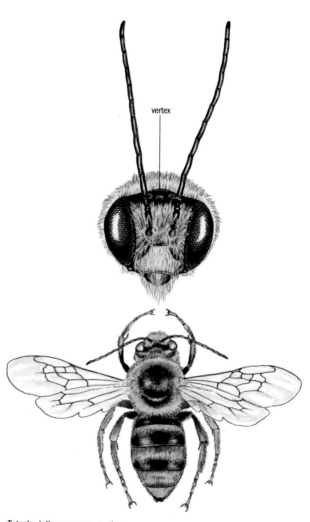

vertex

Tetraloniella pomonae, native

interiors of the fore wing cells have microscopic hairs across the surface. Their abdomen is either covered with pale hair bands or covered completely.

SIMILAR INSECTS: These bees look very similar to bees in the genus *Melissodes* but fly at different times. They also resemble *Eucera* but are generally smaller in size.

FOOD RESOURCES: There is not enough known about these species to be definitive. *T. davidsoni* is considered a specialist on *Encelia* and *Coreopsis*. *T. pomonae* is a specialist on *Hemizonia*.

NESTS: These are ground-nesting bees. Very little is known about this genus in particular though.

FLIGHT SEASON: These are spring bees, though *T. pomonae* is found in the late summer to fall.

CUCKOO BEES   Genus *Triepeolus*

Common California species: *Triepeolus melanarius*   NATIVE

GENUS SUMMARY: *Triepeolus* (tri-e-PEE-o-lus) is a large Holarctic genus. Its members are bee parasites, particularly those of the tribe Eucerini such as *Melissodes* and *Tetraloniella*. There are approximately 150 identified species worldwide, 108 species in North America, and 41 in California, of which 7 are unique to California.

DESCRIPTION: *Triepeolus* are very dark black, medium-size bees, 0.3–0.7 in. (7–17 mm) long, with distinctive white to cream markings on the abdomen. They have no scopal hairs and almost appear beveled. These bees look armored, which reflects their lifestyle as nest invaders. They have three submarginal cells. Under magnification, you can see a prominent central, often-beveled patch of long hairs (setae) on the fifth tergum that are dark and silvery. The patch of hairs is usually as long as it is wide. In males, the mandible is always simple, meaning there are no teeth on the upper margin.

SIMILAR INSECTS: These bees resemble robust wasps. They also look very similar to a closely related genus, *Epeolus*,

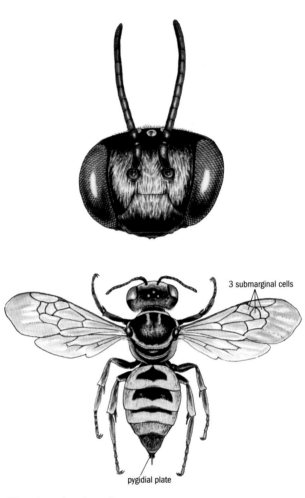

3 submarginal cells

pygidial plate

Triepeolus melanarius, native

which we do not cover in this book. In females the obvious difference between *Epeolus* and *Triepeolus* is the extent and configuration of a hair patch at the tip of the abdomen called the pseudopygidial area (shorter and more transverse in *Epeolus*). These hairs are often silvery.

FOOD RESOURCES: *Triepeolus* visit flowers only to gather nectar. They lay their eggs in the nests of other bees, so they do not need to gather pollen.

NESTS: These bees are cleptoparasitic, so they do not build their own nests. They lay their eggs in the nests of their host bee species. *Triepeolus* have been documented parasitizing bees from almost every family of bees, but are most commonly parasitic on *Melissodes*.

FLIGHT SEASON: These bees match their timing to the bees they parasitize. They are more common in summer to fall, but several species fly in early spring.

APPENDIX 1

BEE FAMILIES AND GENERA FOUND IN CALIFORNIA

The genera we have profiled in this book are identified with an asterisk.

Family Colletidae

Colletes* *Hylaeus**

Family Halictidae

Agapostemon* *Dieunomia* *Nomia*
Augochlora *Dufourea** *Protodufourea*
Augochlorella *Halictus** *Sphecodes**
Augochloropsis *Lasioglossum** *Sphecodosoma*
Conanthalictus *Micralictoides* *Xeralictus*

Family Andrenidae

Ancylandrena *Macrotera** *Perdita**
Andrena* *Megandrena* *Protandrena**
Calliopsis *Panurginus** *Pseudopanurgus*

Family Melittidae

Hesperapis *Melitta*

Family Megachilidae

Anthidiellum *Dianthidium** *Osmia**
Anthidium* *Dioxys* *Protosmia*
Ashmeadiella* *Heriades** *Stelis*
Atoposmia *Hoplitis** *Trachusa*
Chelostoma *Lithurgus* *Xeroheriades*
Coelioxys *Megachile**

Family Apidae

Anthophora*
Anthophorula
Apis*
Bombus*
Brachynomada
Cemolobus
Centris
Ceratina*
Diadasia*
Epeolus
Ericrocis
Eucera*

Exomalopsis
Habropoda*
Hexepeolus
Holcopasites
Leiopodus
Martinapis
Melecta
Melissodes*
Neolarra
Neopasites
Nomada*
Oreopasites

Paranomada
Peponapis*
Rhopalolemma
Svastra
Tetraloniella*
Townsendiella
Triepeolus*
Triopasites
Xenoglossa
Xeromelecta
Xylocopa*
Zacosmia

KEY TO FEMALES OF GENERA INCLUDED IN THIS BOOK

An X indicates that the genus has the character. A blank means the genus does not have the character or that it is variable. Identifying characteristics are visible with a dissecting microscope or hand lens.

Genus	Size		Thorax integument color				Markings	
	Equal to or larger than *Apis*	Smaller than *Apis*	Black or brown	Bright green	Dull metallic black, dark blue, or green	Red	Red	Yellow or white
Andrena	X	X	X		X			X
Osmia	X	X	X		X			
Nomada	X	X	X			X	X	X
Triepeolus	X	X	X				X	
Peponapis	X	X	X					X
Megachile	X	X	X					
Hoplitis	X	X	X					
Halictus	X	X	X					
Anthophora	X	X	X					
Colletes	X	X	X					
Melissodes	X	X	X					
Agapostemon	X	X		X	X			
Bombus	X		X					
Xylocopa	X		X					

Wings	
Number of submarginal cells	Identifying characteristics
3	Wide facial fovea, lined with flat, tiny, white to brown hairs
2	No pale markings on integument; arolium present; parapsidal line condensed to simple pit or short line
2 or 3	Thin and wasplike; most commonly head, thorax, and abdomen bright red with usually extensive yellow markings; with black integument, yellow markings always present; scopa absent; jugal lobe of hind wing small, one-sixth as long as vannal lobe or less
3	No scopal hairs; beveled
3	Clypeus prominent viewed from side; base of mandible dark; end of mandible with distinct notch or tooth at tip; notch relatively shallow, may be gone in older individuals with mandible worn away; interior of fore wing's cells with microscopic hairs scattered across surface
2	No pale markings on integument; arolium absent; scopa on sternites
2	No pale markings on integument; arolium present; hooked apical flagellomeres
3	Basal vein strongly arched at base of vein; bands of hair lining rim of tergites; scopal hairs on tergites
3	Often similar to bumble bee; vertex flattened or slightly concave; interior of fore wing's cells with no microscopic hairs scattered across surface; upper and lower margins of third submarginal cell about equal length; first recurrent vein ending near middle of second submarginal cell
3	Second recurrent vein with lower portion distinctly arched outward toward wing tip, appearing S-shaped
3	Tegula not symmetrical, narrowed anteriorly; clypeus little or moderately bulging in profile; vertex flat or slightly concave; interior cells of wings with minute hairs
3	Bright, shiny, race car green
3	Corbicula on hind tibia present on pollen-carrying species; thorax and abdomen densely hairy; hind wing without jugal lobe
3	Large as queen bumble bee; marginal cell long and narrow; stigma absent; hind tibia without corbicula, with normal scopal hairs

continued →

Genus	Size		Thorax integument color				Markings	
	Equal to or larger than *Apis*	Smaller than *Apis*	Black or brown	Bright green	Dull metallic black, dark blue, or green	Red	Red	Yellow or white
Dufourea		X	?					
Tetraloniella	X		?					
Perdita		X	X		X		X	
Lasioglossum		X	X		X			
Hylaeus		X	X				X	X
Dianthidium		X	X				X	X
Sphecodes		X	X				X	
Anthidium		X	X					X
Panurginus		X	X					
Heriades		X	X					
Ashmeadiella		X	X					
Ceratina		X			X			X
Apis			X					

Wings	
Number of submarginal cells	Identifying characteristics
2	Antennal bases set well below middle of face and separated from clypeus by little more than diameter of antennal socket; clypeus short and wide, its upper margin not much arched up into rest of face; scopal hairs on femur; labrum nearly as long as clypeus
3	Clypeus not strongly bulging or protuberant; vertex flattened to slightly concave; mandible simple without teeth or notches on upper or lower edges; interior of fore wing's cells with microscopic hairs scattered across surface
2	Two subantennal sutures; facial fovea relatively small, narrow, and not filled with hair; short marginal cell
	Basal vein strongly arched at base of vein
2	No scopa; jugal lobe of hind wing about three-fourths as long as vannal lobe; narrow slots or foveae running along compound eye from antennal base to ocelli
2	Yellow, white, or ocher markings on tergites; arolium present; scopa on underside of abdomen; apical end of middle tibia with one small outer spine or sometimes just a tooth
2	Brown to black head and thorax, but almost always with bright red abdomen; basal vein strongly arched at base of vein; no scopal hairs
2	No arolium; yellow or white markings on tergites; subantennal sutures nearly straight; six or more teeth on mandible
2	Two subantennal sutures; facial fovea small, narrow, and not filled with hair; long marginal cell
2	No pale markings on integument; coarse, dense punctuation
2	No pale markings on integument; arolium present; scopa on sternites
3	Shiny; hairs short and sparse throughout with no abdominal bands; clypeus almost always with short white or yellowish bar; end of abdomen forming a point
3	Brown and with sparse pale hairs; eyes with obvious long hairs; corbicula present on hind tibia

GLOSSARY

Abdomen Third segment of an insect.

Antennae Sensory structures that receive and process volatile substances responsible for scent and taste. They also perceive vibrations, air movement, temperature, and humidity. Composed of three segments: scape (base), pedicel (stem), and flagellum.

Anther Male structure where pollen is located on a flower.

Arolia Soft pads between the tarsal claws.

Batesian mimicry Resemblance of a harmless species to a poisonous species that shares a common predator, imitating the signals of the poisonous species.

Carnivorous Meat-eating.

Carpel Ovule- and seed-producing organ in a flowering plant.

Corbicula Pollen basket on the third pair of legs.

Corolla Part containing all of the petals of a flower.

Coxa Part of the leg closest to the abdomen.

Crop Prestomach organ that stores food in an insect.

Exoskeleton Protective covering over the body.

Facial foveae Small depressions or grooves in the face.

Femur Middle part of the leg, past the trochanter and before the tibia.

Fertilization Joining of the sperm and egg.

Filament Structure on which a flower's anther is located.

Flagellomere Subunits of the flagellum.

Flagellum Whiplike end structure of antenna.

Gamete Sex cell.

Gametophyte Plant structure that produces egg and sperm cells.

Glabrous Sparsely hairy.

Hamuli Specialized hooks that connect the two wings together.

Microspore In a plant, the spores that develop into pollen.

Müllerian mimicry Resemblance of two or more poisonous species that share common predators, imitating each other's warning signals.

Nectar Sugar-rich liquid produced by plants.

Nectar guide Pattern or coloration on a flower's petals that directs a bee toward the nectar.

Nectary Plant structure that produces nectar.

Ocelli Simple eyes that register intensity, wavelength, and duration of light.

Ovary Plant structure containing the ovules. It eventually becomes the fruit wall.

Ovipositor Organ used for laying eggs. It also functions as the stinger.

Ovule Structure that contains and produces female reproductive cells in a plant. After fertilization, the ovule becomes the seed.

Pistil Flower organ made up of the stigma, style, and ovary.

Pollen Male gametophytes in a plant.

Propodeum First abdominal segment. It is fused with the thorax.

Pygidial plate Part of the last tergum of a bee. It is often flat and bounded by ridges. It is often triangular in shape.

Scopa Pollen-carrying structure on a bee. The plural is *scopae*.

Spiracles Small holes in the exoskeleton.

Sternum Bottom plate on the abdomen of the bee. The plural is *sterna*.

Stigma Receptive tip of the carpel or of several carpels of a flower that receives the pollen grains.

Style Stalk that pollen tubes grow through to reach the ovary. The stigma is usually on the end of the style.

Subantennal sutures Sutures below the antennae.

Suture Seam between plates in the exoskeleton of an insect, usually seen as a line.

Tarsus End part of the leg, below the tibia.

Tergum Upper plate on the abdomen of a bee. The plural is *terga*.

Thorax Second or middle segment of an insect. The thorax includes the legs and wings.

Tibia Middle part of the leg, below the femur and above the tarsus.

Tibial spur Spiny projection from the tibia.

Trochanter Middle part of the leg, between the coxa and femur.

RESOURCES

Key Readings

Buchmann, Stephen, and Gary Paul Nabhan. 1996. *The Forgotten Pollinators*. Washington, DC: Island Press. The book that started the public interest in pollinators.

Committee on the Status of Pollinators in North America and the National Research Council. 2007. *Status of Pollinators in North America*. Washington, DC: National Academies Press.

Heinrich, Bernd. 2004. *Bumblebee Economics*. Cambridge, MA: Harvard University Press.

Kearns, Carol, and James Thomson. 2001. *The Natural History of Bumblebees*. Boulder: University of Colorado Press. Information on all aspects of North American bumble bee biology as well as instructions for maintaining bumble bees in captivity. The book includes a photographic field guide to aid in the identification of over 50 species of North American bumble bees.

Krombein, Karl V., Paul D. Hurd Jr., David R. Smith, and B. D. Burks. 1979. *Catalog of Hymenoptera in America North of Mexico*. Washington, DC: Smithsonian Institution Press. Available online.

Michener, Charles D. 2007. *The Bees of the World,* 2nd ed. Baltimore: The Johns Hopkins University Press. The definitive academic text on bee biology by the dean of bee biology.

Michener, Charles D., Ronald J. McGinley, and Bryan N. Danforth. 1994. *The Bee Genera of North and Central America (Hymenoptera: Apoidea)*. Washington, DC: Smithsonian Books. Technical scientific key to the genera found in North America.

O'Toole, Christopher, and Anthony Raw. 2004. *Bees of the World*. New York: Facts on File. A wonderful book, even if the publisher put a fly on the cover.

Stephen, W. P., G. E. Bohart, and P. F. Torchio. 1969. *The Biology and External Morphology of Bees: With a Synopsis of the Genera of Northwestern America*. Corvallis: Agricultural Experiment Station, Oregon State University-Corvallis. A readable though technical guide with detailed discussions of bee morphology, behavior, and biology.

Thorp, Robbin, Donald S. Horning Jr., and Lorry L. Dunning. 1983. *Bumble Bees and Cuckoo Bumble Bees of California*. Berkeley: University of California Press.

Xerces Society. 2011. *Attracting Native Pollinators*. North Adams, MA: Storey Press.

Key Web Resources

www.bugguide.com
Website for posting images of bugs. Some of the best photographs of insects can be found here. The site includes images of many bees that have been identified to the species level.

www.discoverlife.org
Online keys for many different plants and animals including bees. The key to genera for bees is excellent.

www.fs.fed.us/wildflowers/pollinators
US Forest Service guide to pollinators.

www.knoxcellars.com
Commercial source for bee houses.

http://nature.berkeley.edu/urbanbeegardens/
Dr. Gordon Frankie's wonderful urban bee gardening site.

www.pollinator.org
Bee articles and planting and gardening guides from Pollinator Partnership.

www.xerces.org
The premier invertebrate conservation group's site with pollinator fact sheets and planting guides.

INDEX

ABOUT THE AUTHOR
AND ILLUSTRATOR

Gretchen LeBuhn is a professor of Biology at San Francisco State University and Director of the Great Sunflower Project.

Noel B. Pugh received a BFA in Painting from the San Francisco Art Institute and a Masters Certificate in Science Illustration from UC Santa Cruz. He is excited to count this as his first important publication in the field of scientific illustration. The majority of these illustrations were completed by observing pinned specimens in the laboratory of Gretchen LeBuhn at San Francisco State University and the California Academy of Sciences. You can view more of his artwork at www.fullpollenbasket.com.

California Natural History Guides

From deserts and grasslands to glaciers and the spectacular Pacific coast, California is a naturalist's paradise. The authors of the California Natural History Guides have walked into wildfires, plunged into shark-infested waters, scaled the Sierra Nevada, peeked under rocks, and gazed into the sky to present the most extensive California environmental education series in existence. Packed with photographs and illustrations and compact enough to take on the trail, they are essential reading for any California adventurer. For a complete list of guides, please visit http://www.ucpress.edu/go/cnhg.